Microstructure Atlas of P22 Steel

Rajat K. Roy · Anil K. Das · Avijit Kumar Metya · Avijit Mondal · Ashis Kumar Panda · M. Ghosh · Satish Chand · Sarmishtha Palit Sagar · Swapan Kumar Das · Amit Chhabra · Swaminathan Jaganathan · Amitava Mitra

Microstructure Atlas of P22 Steel

CSIR
भारत का नवाचार इंजन
The Innovation Engine of India

एनटीपीसी
NTPC

Springer

Rajat K. Roy
CSIR-National Metallurgical Laboratory
Jamshedpur, Jharkhand, India

Avijit Kumar Metya
CSIR-National Metallurgical Laboratory
Jamshedpur, Jharkhand, India

Ashis Kumar Panda
CSIR-National Metallurgical Laboratory
Jamshedpur, Jharkhand, India

Satish Chand
NTPC Energy Technology Research Alliance
Greater Noida, Uttar Pradesh, India

Swapan Kumar Das
CSIR-National Metallurgical Laboratory
Jamshedpur, Jharkhand, India

Swaminathan Jaganathan
CSIR-National Metallurgical Laboratory
Jamshedpur, Jharkhand, India

Anil K. Das
NTPC Energy Technology Research Alliance
Greater Noida, Uttar Pradesh, India

Avijit Mondal
NTPC Energy Technology Research Alliance
Greater Noida, Uttar Pradesh, India

M. Ghosh
CSIR-National Metallurgical Laboratory
Jamshedpur, Jharkhand, India

Sarmishtha Palit Sagar
CSIR-National Metallurgical Laboratory
Jamshedpur, Jharkhand, India

Amit Chhabra
NTPC Energy Technology Research Alliance
Greater Noida, Uttar Pradesh, India

Amitava Mitra
CSIR-National Metallurgical Laboratory
Jamshedpur, Jharkhand, India

ISBN 978-981-99-0116-6 ISBN 978-981-99-0117-3 (eBook)
https://doi.org/10.1007/978-981-99-0117-3

This Springer imprint is published by the registered company Springer Nature Singapore Pte Ltd.
The registered company address is: 152 Beach Road, #21-01/04 Gateway East, Singapore 189721, Singapore

This work is prepared by CSIR-National Metallurgical Laboratory, Jamshedpur, India; in-collaboration with NTPC-NETRA, Greater Noida, India

Acknowledgements

The authors take this opportunity to express their gratitude to heads of CSIR-NML, IPU-CSIR and NTPC-NETRA for their kind permission to publish this research work and wish to record their heartfelt thanks to Shri Ujjwal Kanti Bhattacharya, Director (Project), NTPC Ltd. and Mr. Shaswattam, Chief General Manager (CGM), NTPC-NETRA, for their continuous guidance, active interest, untiring co-operation and financial support to execute the proposed investigation.

Special thanks are extended to Mr. Binod Kalindi and Mr. Ankur Chatterjee for mechanical tests, Mr. Anuj Macchua and Mr. Sadashiv Kalindi during metallographic investigation, Dr. Bhupeswar Mahato for TEM investigation and Dr. Siuli Dutta for scanning electron microscopy during the course of investigation. Finally, the unconditional support from other staff members of CSIR-NML at various stages of investigation in the needy hours is also gratefully acknowledged. The team is thankful to NTPC Sites (Kahalgaon and Barh) for providing the samples of P22 for the experimentations.

Contents

1 **Metallurgy of P22 Steel** 1
References 3

2 **As-Received Pipe Details** 5
2.1 Dimensions: Pipe and Creep Specimen 5
2.2 Material Composition 6
2.3 Mechanical Properties 6

3 **Creep Testing Under Different Conditions** 7
3.1 Creep Life of P22 Steel 7
3.2 Creep Curves 8

4 **Micrographs Description** 11
4.1 As-Received Pipe 11
4.2 Creep Exposure at 550 °C Temperature and 115 MPa Stress 16
4.3 Creep Exposure at 575 °C Temperature and 115 MPa Stress 50
4.4 Creep Exposure at 600 °C Temperature and 100 MPa Stress 72

5 **Effect of Creep Exposure on Material Behaviours** 93
5.1 Precipitate Size Variation 93
5.2 Hardness Variation 93
5.3 Magnetic Coercivity and MBE Voltage 93
5.4 Nonlinear Ultrasonic (NLU) Parameter 94
5.5 Conventional Ultrasonic Parameter 94
Supplements 103

About the Authors

Dr. Rajat K. Roy is currently positioned as a Senior Principal Scientist at CSIR-National Metallurgical Laboratory, Jamshedpur (India). After completing his Bachelor of Engineering and Master of Engineering in Metallurgy from Bengal Engineering College, Shibpur, Howrah (India), he obtained his Ph.D. degree in Metallurgical and Materials Engineering from Indian Institute of Technology, Kharagpur (India). His research areas are compositional design and development of nanostructured and amorphous materials through rapid solidification, metal joining (brazing), characterization of ferritic steels by electromagnetic NDE. He has authorships of 54 papers in SCI journals, 12 patents, 3 copyrights and 2 book chapters. Dr. Roy has exposures on the development of nanomaterials in Carnegie Mellon University, Pittsburgh (USA) under BOYSCAST fellowship and Korea University, Seoul (Republic of Korea) under BK21 program.

Mr. Anil K. Das is working as General Manager at NTPC Energy Technology Research Alliance (NETRA), the R&Dwing of NTPC Limited (India).He is heading the Advanced Scientific Service Division there, which has labs/group related to Non-destructive Evaluation (NDE), Advanced Material Research, Creep & Fatigue, Corrosion and Environment, Analytical Chemistry, Lub & Transformer Oils, Coal & Combustion and Resin. He has done his B.Tech and M.Tech both from I.I.T. Kanpur (India). He has more than 30 years of experience in health and life assessment of power plant components. He was awarded National NDT Award in 2011 for Excellence in Application of NDE in Industries. He has filed 5 patents (2 granted) and 5 copyrights. He is life member of ISNT and member of ASNT. He is ASNT Level-3 certified in four NDT Methods (UT, MT, ET & PT) and CSWIP-Level 2 in PAUT. He is associated with more than 350 publications of reports and technical papers. His major work areas are advanced inspection and analysis tools for power plant components.

Dr. Avijit Kumar Metya has been working at CSIR-NML, Jamshedpur (India) as Principal Scientist. He obtained his B.E. (Mechanical) from Jalpaiguri Government Engineering College, West Bengal (India); M.Tech. in Materials Engineering from BESU, Shibpur (currently IIEST, Shibpur), Howrah (India) and Ph.D. from Indian Institute of Technology Madras, Chennai (India). His major research interests include Ultrasonic NDE, acoustic wave propagation through materials, mechanical behaviour of materials. He has published 16 papers in International Journals. Dr. Metya received NDT Young Scientist/Engineer Award in 2011 from Indian Society for Nondestructive Testing.

Dr. Avijit Mondal is currently working as Senior Manager at NETRA, India. He obtained his B.E and M.Tech. from Jadavpur University, Kolkata (India) and Ph.D. from the Indian Institute of Technology, Kanpur (India). His major research area includes structure properties correlation, advanced NDE methods for condition assessment, failure analysis of power plant components, etc. Before joining NTPC, he was working as visiting Research Scholar at the Pennsylvania State University, USA. He was awarded Shunsaku Katoh Award for most significant contributions to scientific research in the field of Microwave and RF power at the 1st Global Congress on Microwave Energy Applications (GCEMA 2008), Japan. He was selected as one of the two participants from IIT Kanpur for SAARC-JENESYS 2009 Program by the Japanese Government. He has authored about 32 publications in international journals, conference proceedings and books. He has filed 2 patents and 4 copyrights.

Dr. Ashis Kumar Panda is working as Senior Principal Scientist in the Functional Materials (FM) group of Advanced Materials and Processes (AMP) Division of National Metallurgical laboratory, Jamshedpur (India). He obtained his Ph.D. degree in Materials Science from Indian Institute of Technology, Kharagpur (India) in the year 2002. His research interest is in the development of magnetic alloys (nanostructured, ferromagnetic shape memory, magnetocaloric) through rapid solidification, non-destructive evaluation through magnetic techniques and development of magnetic sensors. He has research experience in Slovak Academy of Science (SAS), Bratislava, Slovakia and National Institute of Materials Science (NIMS), Tsukuba, Japan under Indian National Science Academy (INSA) visiting program and CSIR-Raman Research fellowship, respectively. He has about 96 publications in SCI journals and has filed 18 patents.

Dr. M. Ghosh is at present Senior Principal Scientist in Microstructural Engineering Group of Materials Engineering Division, CSIR-National Metallurgical Laboratory, Jamshedpur 831007 (India). After completing B.E. in Metallurgy from Bengal Engineering College, Shibpur, Howrah 711103 (India) and M.E. in Physical Metallurgy from BE College (Deemed University), Howrah 711103 (India), he obtained his Ph.D. in Engineering from BESU, Howrah 711103 (India). Dr. Ghosh was awarded with Indo-US Fellowship in the year of 2009 for post-doctoral research in abroad on advanced joining protocols. He has more than seventeen years of experience in research and development with a major thrust in the area of structure-property co-relation of metallic materials, transmission electron microscopy, solid state joining and failure investigation of engineering components. He has published more than 115 papers in the journals of international repute along with 15 conference proceedings.

Mr. Satish Chand is currently working as Manager at NTPC Energy Technology Research Alliance (NETRA), India. He obtained his M.Sc. (Physics) from Meerut University (India). He has more than 25 years of experience in the field of health and life assessment of power plant components. His major research area includes developing procedures and methodologies for NDE inspection of power plant components, failure investigation, providing new and innovative solutions. He has published more than 15 papers in in-house journal and in international conferences. He has filed 2 patents and 2 copyrights.

Dr. Sarmishtha Palit Sagar obtained her Master of Technology degree in Electronics and Communication Engineering from University of Calcutta, Kolkata, India in 1993 and her Ph.D. from Jadavpur University, Kolkata, India in 2003. She completed her post-doctoral studies at Fraunhofer Institute for Non-Destructive Testing (IzFP), Saarbreucken, Germany and Pennsylvania State University, USA. She currently holds the position of Chief Scientist and Group Leader, NDE group in CSIR-National Metallurgical Laboratory, Jamshedpur, India. She has published 60 SCI/non-SCI journal papers, 13 patents and 06 copyrights. Her current research interest includes Non-Destructive Evaluation using Ultrasonic, Thermography and Advanced Ultrasonic, Sensor and devices for industrial applications, Signal analysis, Internet of Things (IoT) and Artificial Intelligence (AI). She is an editorial board member of the Journal of Non-destructive Evaluation. Dr. Palit Sagar received "National NDT Award in R&D" in 2005, "National NDT award for international recognition" in 2017 and inducted as Honorary fellow of Indian Society for Non-destructive Testing (ISNT).

Dr. Swapan Kumar Das Ex-Senior Principal Scientist of CSIR-National Metallurgical Laboratory, and Professor of AcSIR was superannuated in February 2020 after 30 years of service. He obtained his M.Engg. and Ph.D. in Metallurgical and Materials Engineering from IIEST, Shibpur, Howrah 711103 (India). His areas of specialization are Physical Metallurgy, Materials Characterization and Failure Analysis. He has published more than 100 papers in reputed SCI journals.

Mr. Amit Chhabra is currently Senior Manager at the Mechanical Section, NTPC Khargone (India). Earlier, he worked in NDE, Life Management Laboratory at NTPC Energy Technology Research Alliance (NETRA, R&D wing of NTPC). He obtained his B.Tech. (Mechanical) from National Institute of Technology, Jaipur (India). His major work areas at NETRA include developing procedures and methodologies for NDE inspection of power plant components, failure investigation, providing new and innovative solutions. He is CSWIP Level 2 certified in PAUT & ToFD. He has published more than 10 papers in in-house journal and in international conferences. He has filed 2 patents and 2 copyrights.

Dr. Swaminathan Jaganathan is currently working in the position of Chief Scientist at Creep Lab., CSIR-National Metallurgical Laboratory, Jamshedpur (India). He has 28 years research experiences on creep deformation of high temperature alloys. His research interests are remaining life assessment of high temperature components and failure analysis of industrial steel components. He has published about 35 nos. research papers in various reputed journals.

Prof. Amitava Mitra is currently Visiting Professor at Department of Physics and Professor In-charge (Infrastructure), Indian Institute of Technology Jodhpur, Rajasthan (India). He did his Ph.D. from Indian Institute of Technology, Kharagpur (India) and worked about 30 years at CSIR-National Metallurgical Laboratory, Jamshedpur (India). His major areas of research interest include Magnetism and Magnetic Materials and their applications in the field of structural integrity assessment of engineering components. He published more than 150 papers in SCI journals and filed about 15 patents in India and abroad. He is the recipient of National Metallurgist Day Award: Metallurgist of the year-2006 conferred by Ministry of Steel, Government of India for his contributions in the field of Metal Science. He is also the recipient of National NDT award in the year 1997 for R&D and in the year 2000 for system development which were conferred by Indian Society for NDT. He received the Materials Research Society of India Medal in the year 2011. He is an USAID Fellow and worked at Ames Laboratory, Iowa state University, USA and also a JSPS invitation Fellow and worked at Institute of Materials Research, Tohoku University, Sendai, Japan and Iwate University, Morioka, Japan. He is a Fellow of Indian Institute of Metals, Institute of Engineers (India) and Indian Society for NDT.

Chapter 1
Metallurgy of P22 Steel

Two major parameters affect the thermal efficiency of a power plant: one of them is temperature and another is pressure of the steam to operate the turbine. The thermal efficiency of a power plant can be increased significantly by increasing these two parameters. When steam temperature <600 °C with pressures at 25–30 MPa, thermal efficiencies can reach about 42% in most modern steam power stations. This can be increased to around 45% in the next generation of power stations with steam temperatures of 625–650 °C. But the increase in operating temperatures and pressures accelerate all degradation processes in steels, such as recovery and recrystallization, change of nature of dislocations, precipitation of particles, grain-subgrain-second phase coarsening and appearance of other second phases. The structural alteration is also governed by long-term service exploitation.

In conventional thermal power plant, high Cr ferritic steel finds widespread application for temperature <600 °C owing to its attractive mechanical properties, satisfactory fabricability, sufficient oxidation resistance and adequate weldability [1]. In this respect, the austenitic alloys exhibit better high-temperature mechanical strength; however, suffers from thermal fatigue, stress corrosion cracking and high co-efficient of thermal expansion [2].

Out of wide varieties in Cr–Mo steel, the ASTM A335 Grade P22 (2.25Cr–1Mo) steel is one of the widely used materials for the construction of structural and pressure vessel components of thermal power plant, including high-temperature header and piping, superheater tubing, reheater tubing, etc. As per ASME Boiler and Pressure Vessel Code (BPVC) Section 2A, the typical chemical composition of SA335 P22 steel is 0.05–0.15 C, 1.90–2.60 Cr, 0.87–1.13 Mo, 0.50 max Si, 0.30–0.60 Mn, 0.025 max S, 0.025 max P, rest Fe (all in wt%). It obtains a suitable high-temperature strength due to dispersion of fine carbide precipitates in a ferrite matrix at the time of processing by seamless pipe technology. The fractions of matrix phases and carbide precipitates are dependent on heat treatment schedule and subsequently alters with state of stress and temperature, i.e. creep condition as mentioned above during service. The P22 steels are used as annealed or normalized-and-tempered condition, depending on its place of application. Due to different cooling rate, the

R. K. Roy et al., *Microstructure Atlas of P22 Steel*,
https://doi.org/10.1007/978-981-99-0117-3_1

annealed steel comprises ferrite-pearlite aggregate and the normalized-tempered steel evolves ferrite-bainite aggregate. The presence of bainite improves creep resistance at high stress and short time conditions. The bainitic structure in the steel deteriorates rapidly at higher degree than that of the steel with pearlitic structures. Accordingly, the annealed steel shows better intermediate low stress creep resistance [3]. The high-temperature creep exposure results in the diversion from true lamellar pearlite to the discrete pearlite structure, resembling the discontinuous distribution of lamellar pearlitic morphology.

During service exploitation, the creep deformation becomes dominant at temperature of $\geq 0.45\ T_m$ (where T_m = alloy melting point). For P22 steel the same is ~450 °C, which means at normal operating temperature for all power plant materials creep is significant. Apart from matrix microstructure as stated before, the creep strength of Cr–Mo steels is achieved through the solid solution strengthening of ferrite matrix by carbon, chromium and molybdenum, and is further enhanced by carbide precipitation. During service, the change in phase constituents is significant and the $M_{23}C_6$ (where M = Fe, Cr, Mo) phase, occurs in a large volume fraction within matrix. It precipitates at the prior austenite grain boundaries (PAGB), packet boundaries, block boundaries and lath boundaries. It coarsens quickly, reducing the effect of strengthening. Similarly, the M_2C precipitates grow fast.

To summarize the above illustration, the carbides in Cr–Mo steel thus can be categorized in five types, viz., M_3C, M_2C, M_7C_3, $M_{23}C_6$, M_6C (M = Fe, Cr, Mo), according to their morphology, metallic element content and sequence of precipitation (Fig. 1.1).

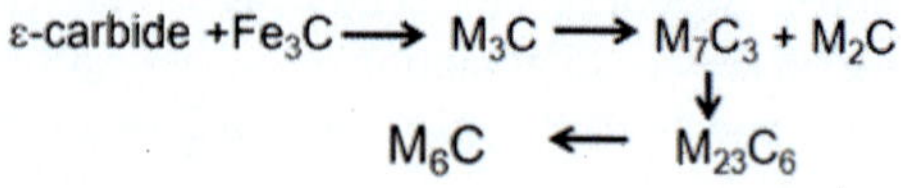

Fig. 1.1 Carbide precipitation reaction sequence in 2.25Cr–1Mo steel during ageing at 750 °C [4]

The material performance in a given condition completely depends on the relative fraction of different phases, their morphology, nature and distribution pattern. The progressive change in microstructure determines the degree of degeneration and the residual life of the component up to its designated tenure of service. Necessity therefore arises to develop a data bank, regarding the qualitative/quantitative structural changes for the mentioned alloy under different state of stress and temperatures to identify exact interval for permissible overhauling and avoid catastrophic failure during service. In the present endeavour, the different structural features that are evolved for P22 alloy under various stress and temperature have been illustrated.

References

1. B. Kuhn, M. Talik, L. Niewolak, J. Zurek, H. Hattendorf, P.J. Ennis, W.J. Quadakkers, T. Beck, L. Singheiser, Development of high chromium ferritic steels strengthened by intermetallic phases. Mats. Sci. Eng. A. **594**, 372–380 (2014)
2. J.A. Francis, W. Mazur, H.K.D.H. Bhadeshia, Type IV cracking in ferritic power plant steels. Mats. Sci. Tech. **22**, 1387–1394 (2006)
3. R. Viswanathan, *Damage Mechanisms and Life Assessment of High Temperature Components* (ASM International, Metal Parks, OH, 1989)
4. S. Saroja, P. Parameswaran, M. Vijayalakshmi, V.S. Raghunathan, Prediction of microstructural states in Cr–Mo steels using phase evolution diagrams. Acta Metall. Mater. **43**, 2985–3000 (1995)

Chapter 2
As-Received Pipe Details

2.1 Dimensions: Pipe and Creep Specimen

See Fig. 2.1.

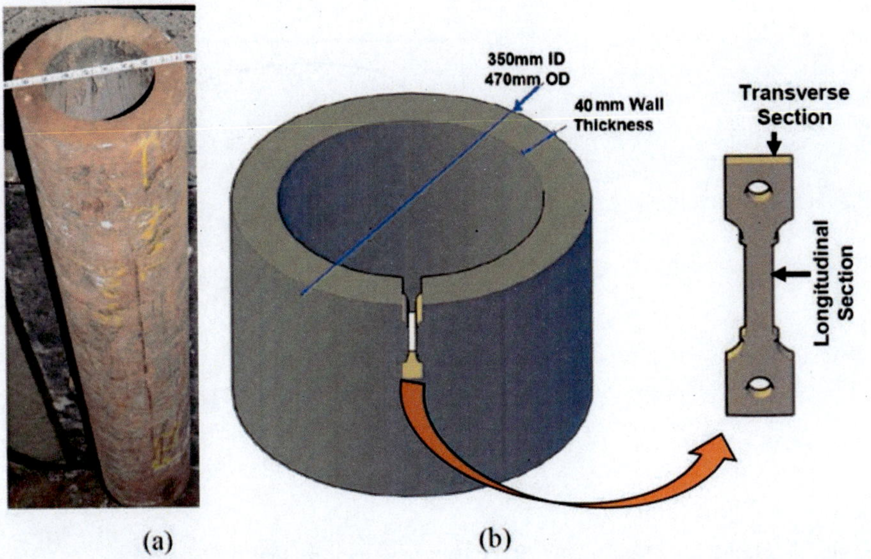

Fig. 2.1 **a** Industrially processed as-received pipe, **b** schematic diagram of creep specimen and sectioning position from pipe for metallography and hardness measurement

R. K. Roy et al., *Microstructure Atlas of P22 Steel*,
https://doi.org/10.1007/978-981-99-0117-3_2

Table 2.1 Chemical composition (wt %) of P22 steel

C	Si	Mn	P	S	Cr	Mo	Ni	Fe
0.135	0.235	0.47	0.012	0.005	2.185	0.93	0.08	Rest

2.2 Material Composition

See Table 2.1.

2.3 Mechanical Properties

See Fig. 2.2 and Table 2.2.

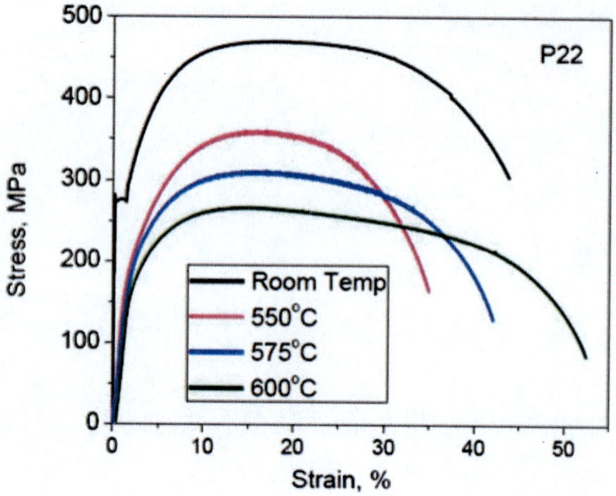

Fig. 2.2 Stress-strain curves of P22 (2.25Cr–1Mo) steel at room and high temperatures

Table 2.2 Mechanical properties of P22 at room temperature and high temperatures

Temp. (°C)	Hardness (HV30)	Fracture toughness , $J_{0.2}$ (kJ/m^2)	Tensile properties		
			Yield strength (MPa)	UTS (MPa)	Fracture strain (%)
Room temp.	132.6 ± 0.6	713.9	277 ± 1.5	465 ± 4.2	38 ± 3.0
550	–	–	166 ± 1.0	343 ± 7.7	20 ± 7.0
575	–	–	154 ± 1.2	295 ± 7.4	26 ± 8.0
600	–	–	133 ± 1.9	234 ± 16.2	34 ± 9.3

Chapter 3
Creep Testing Under Different Conditions

3.1 Creep Life of P22 Steel

Creep life of P22 steel is determined at different temperatures and stresses, as mentioned in Table 3.1. It is noteworthy that the creep life varies for different samples at same temperature and stress condition. The corresponding creep curves are explained in Figs. 3.1a, 3.2a and 3.3a, respectively.

Table 3.1 Creep life of P22 at different conditions

Creep conditions (Temp./Stress)	Sample no	Creep life (h)
550 °C/115 MPa	550(1)	1605
	550(2)	1240
	550(3)	1164
575 °C/115 MPa	575(1)	355
	575(2)	368
	575(3)	315
	575(4)	300
	575(5)	387
600 °C/100 MPa	600(1)	260
	600(2)	200
	600(3)	233

R. K. Roy et al., *Microstructure Atlas of P22 Steel*,
https://doi.org/10.1007/978-981-99-0117-3_3

3.2 Creep Curves

See Figs. 3.1, 3.2 and 3.3.

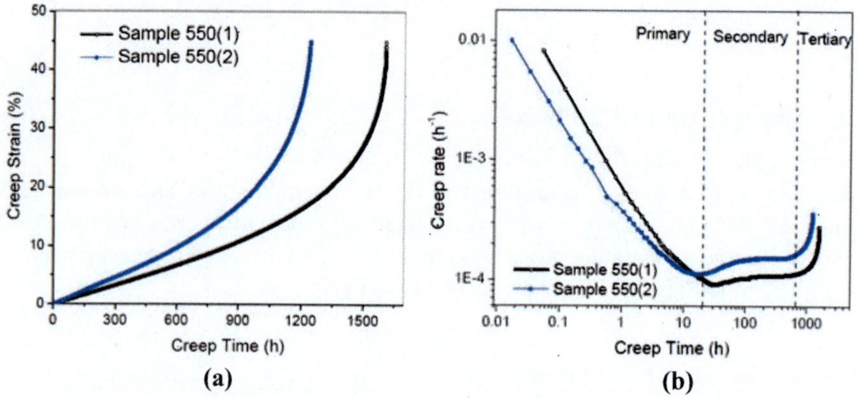

Fig. 3.1 Creep properties of P22 at 550 °C under 115 MPa stress. **a** Creep strain and **b** Creep strain rate, as a function of time

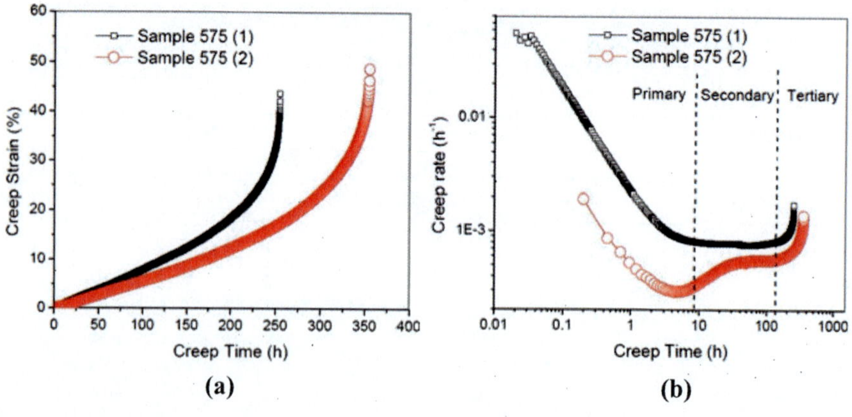

Fig. 3.2 Creep properties of P22 at 575 °C under 115 MPa stress. **a** Creep strain and **b** Creep strain rate, as a function of time

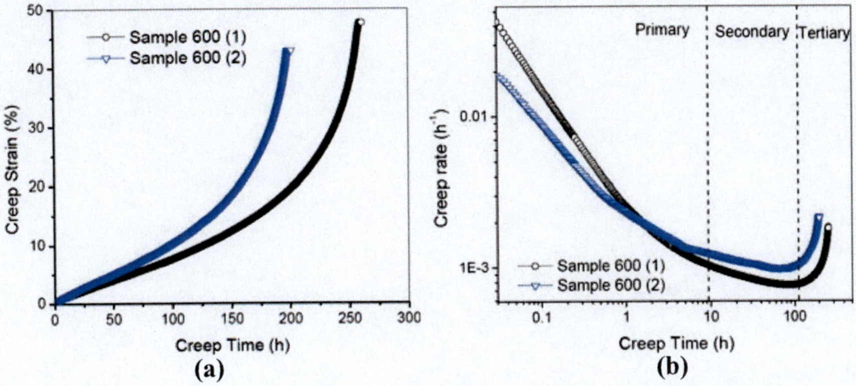

Fig. 3.3 Creep properties of P22 at 600 °C under 100 MPa stress. **a** Creep strain and **b** Creep strain rate, as a function of time

Chapter 4
Micrographs Description

4.1 As-Received Pipe

See Figs. 4.1, 4.2, 4.3 and 4.4.

R. K. Roy et al., *Microstructure Atlas of P22 Steel*,
https://doi.org/10.1007/978-981-99-0117-3_4

Mag. 200X

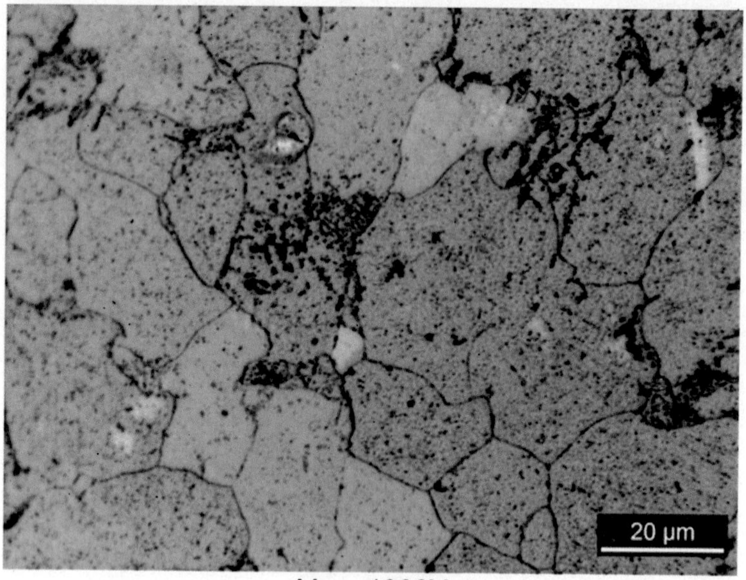

Mag. 1000X

Fig. 4.1 Optical micrographs of as-received P22 (2.25Cr–1Mo) steel at transverse section showing inhomogeneous distribution of pearlite (P) in ferrite (F) matrix

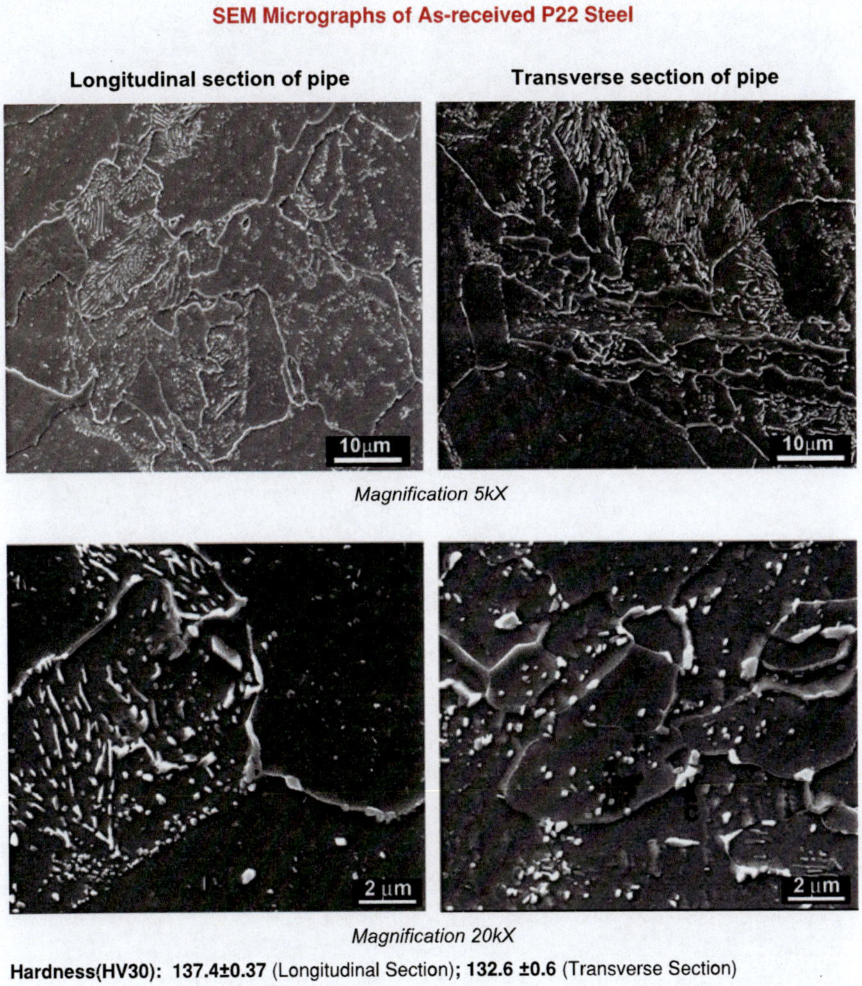

Fig. 4.2 SEM-SE micrographs of as-received P22 steel, showing distribution of carbide (C) precipitates at grain boundary and grain interior of ferrite (F)-pearlite (P) matrix

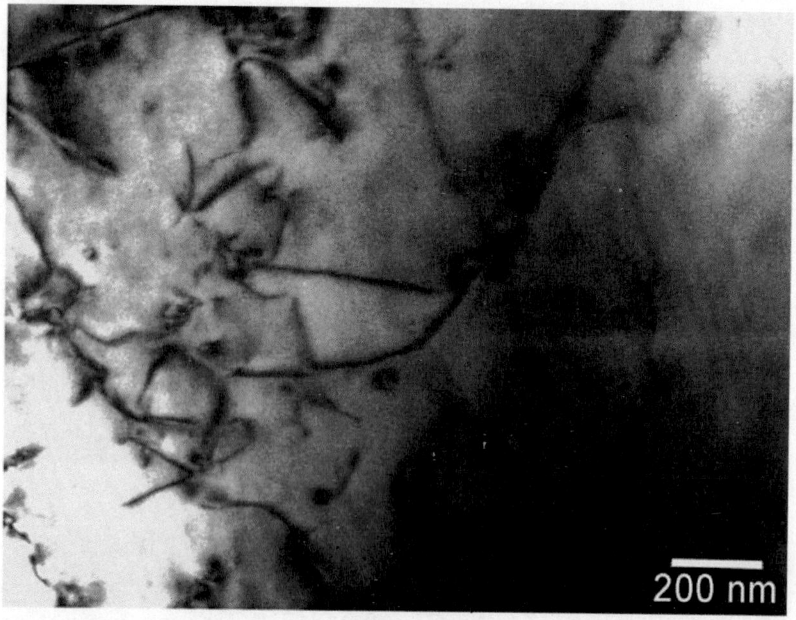

(a), *Mag. 15kX*

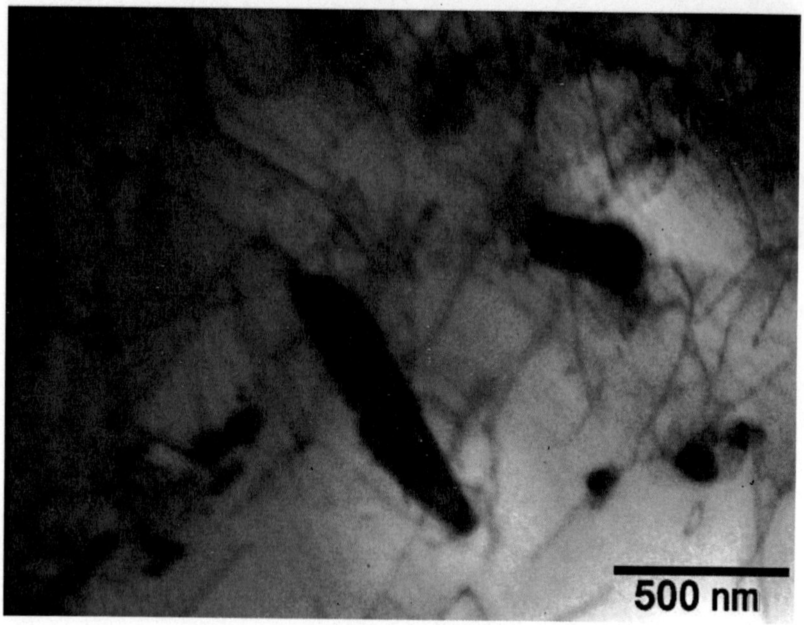

(b), *Mag. 11.5kX*

Fig. 4.3 TEM-BF images of P22 steel as-received condition, **a** dislocation networks, **b** dislocation pinning by large precipitate of (Fe, Cr, Mo)-rich carbides

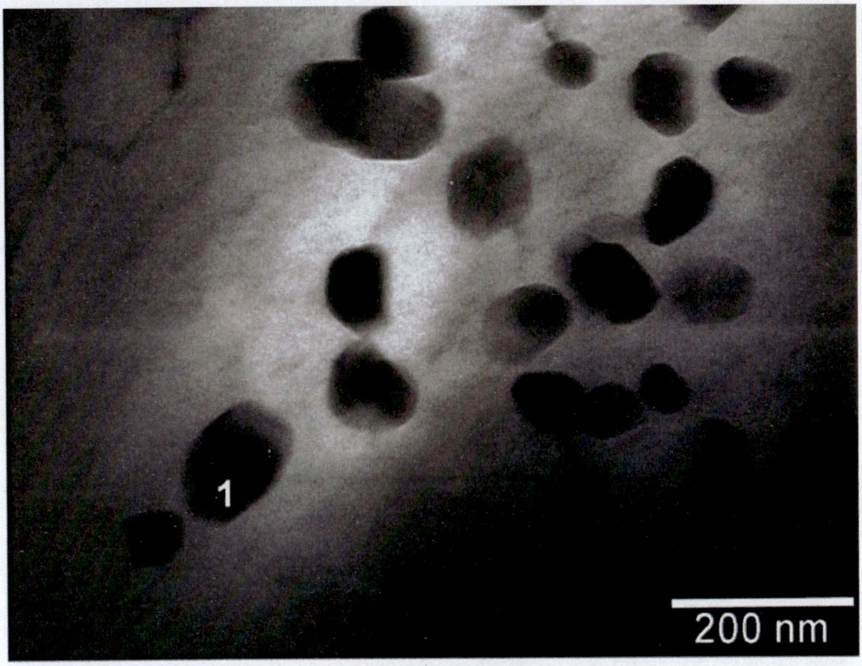

(a), *Mag. 27.5kX*

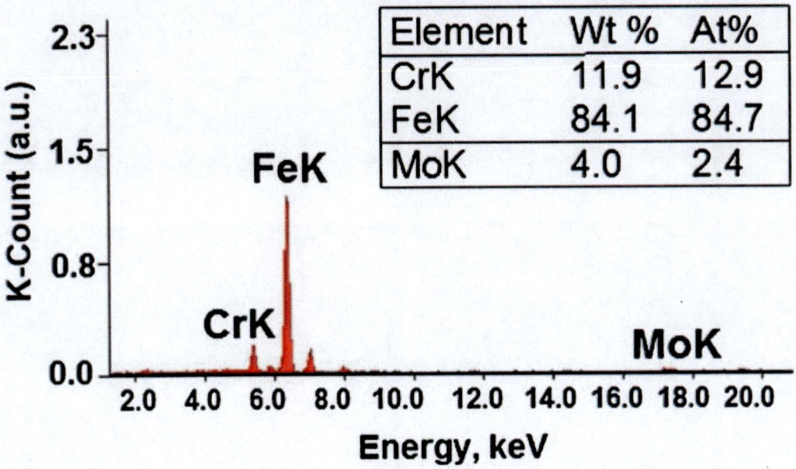

(b)

Fig. 4.4 TEM-BF image of P22 steel as-received condition showing, **a** inhomogeneous distribution of polygonal Cr–Mo precipitate in matrix, **b** EDS analysis of precipitate (1) of this figure (**a**)

4.2 Creep Exposure at 550 °C Temperature and 115 MPa Stress

See Figs. 4.5, 4.6, 4.7, 4.8, 4.9, 4.10, 4.11, 4.12, 4.13, 4.14, 4.15, 4.16, 4.17, 4.18, 4.19, 4.20, 4.21, 4.22, 4.23, 4.24, 4.25, 4.26, 4.27, 4.28, 4.29, 4.30, 4.31, 4.32, 4.33, 4.34, 4.35, 4.36, 4.37, 4.38, 4.39 and 4.40.

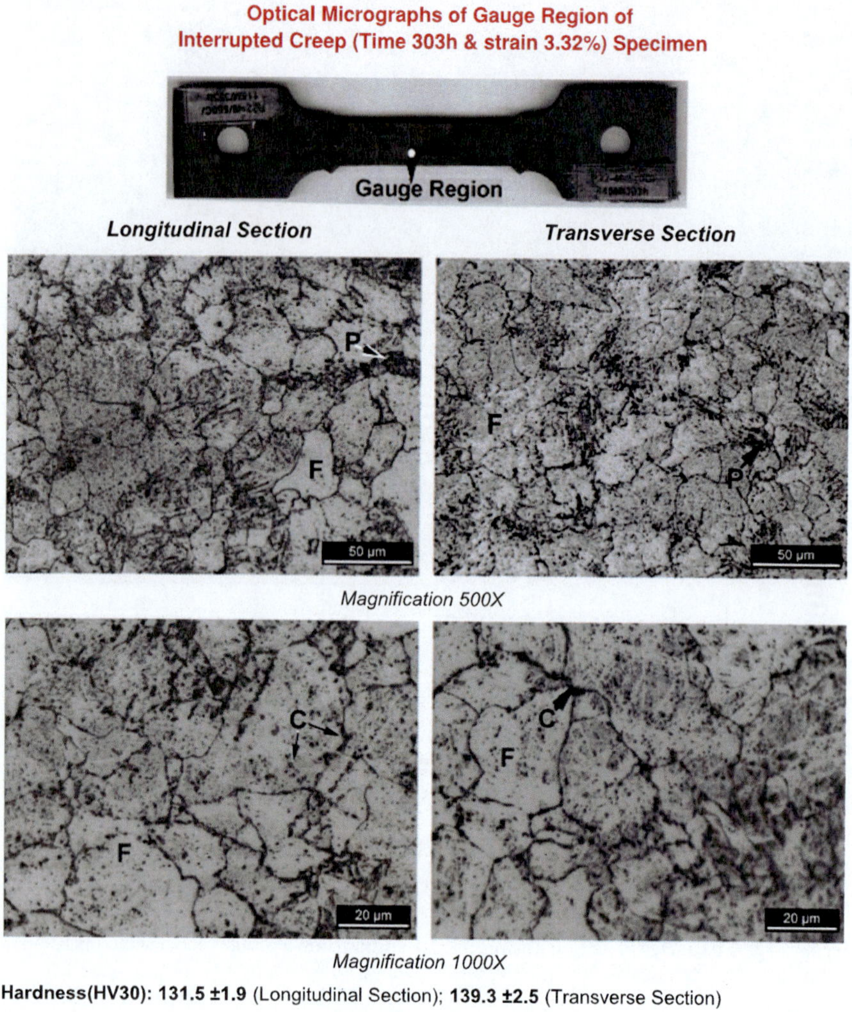

Fig. 4.5 Optical micrographs of **longitudinal and transverse sections** of P22 (2.25Cr–1Mo) steel after creep exposed at 550 °C/115 MPa for 303 h, showing inhomogeneous distribution of pearlite (P) in ferrite (F) matrix and carbides (C) at grain boundaries and grain interiors

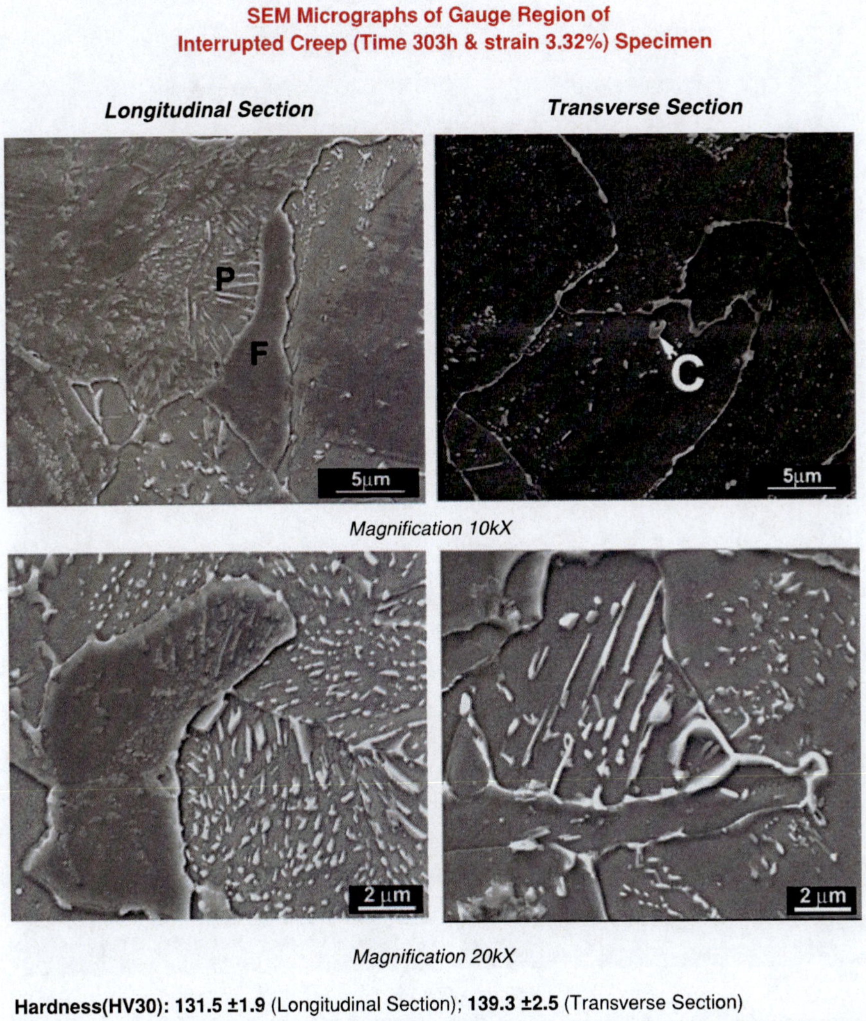

Fig. 4.6 SEM-SE micrographs of **longitudinal and transverse sections** of P22 (2.25Cr–1Mo) steel after creep exposed at 550 °C/115 MPa for 303 h, showing the distribution of (Fe, Cr, Mo)-rich carbides (C) at grain interiors of ferrite (F) and pearlite (P) matrix

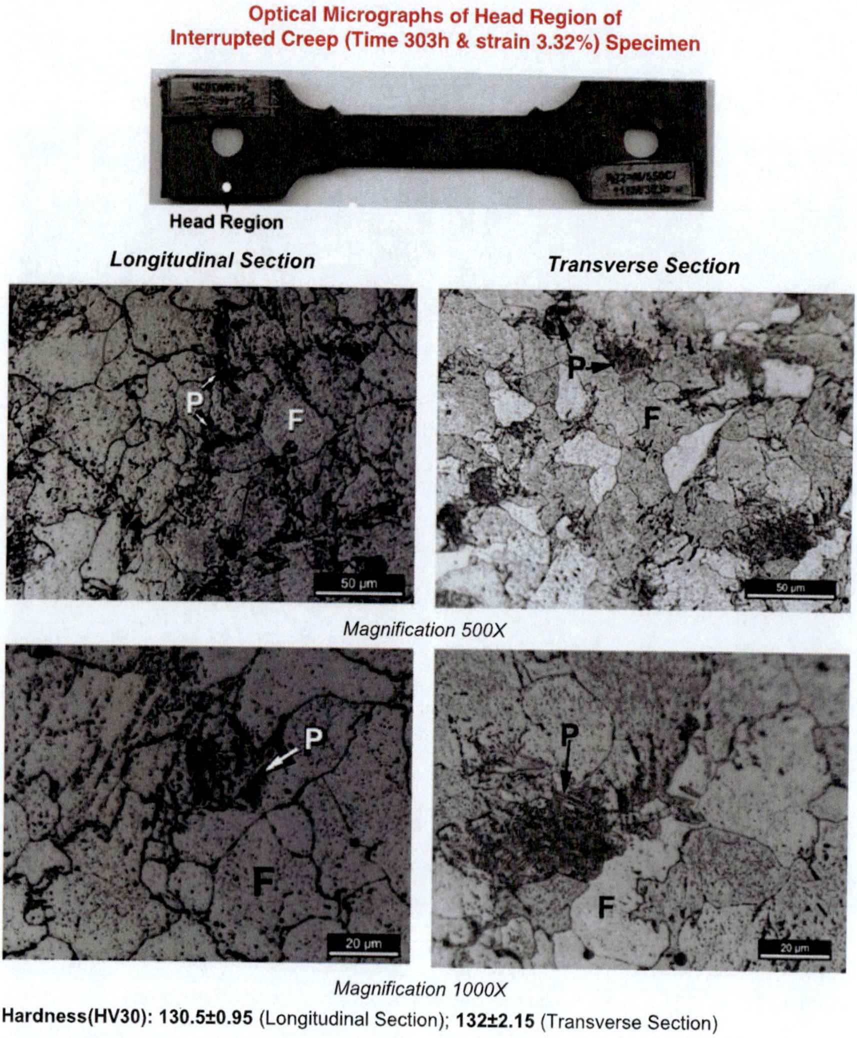

Fig. 4.7 Optical micrographs of **longitudinal and transverse sections** of P22 (2.25Cr–1Mo) steel after creep exposed at 550 °C/115 MPa for 303 h, showing inhomogeneous distribution of ferrite (F) and pearlite (P) structures

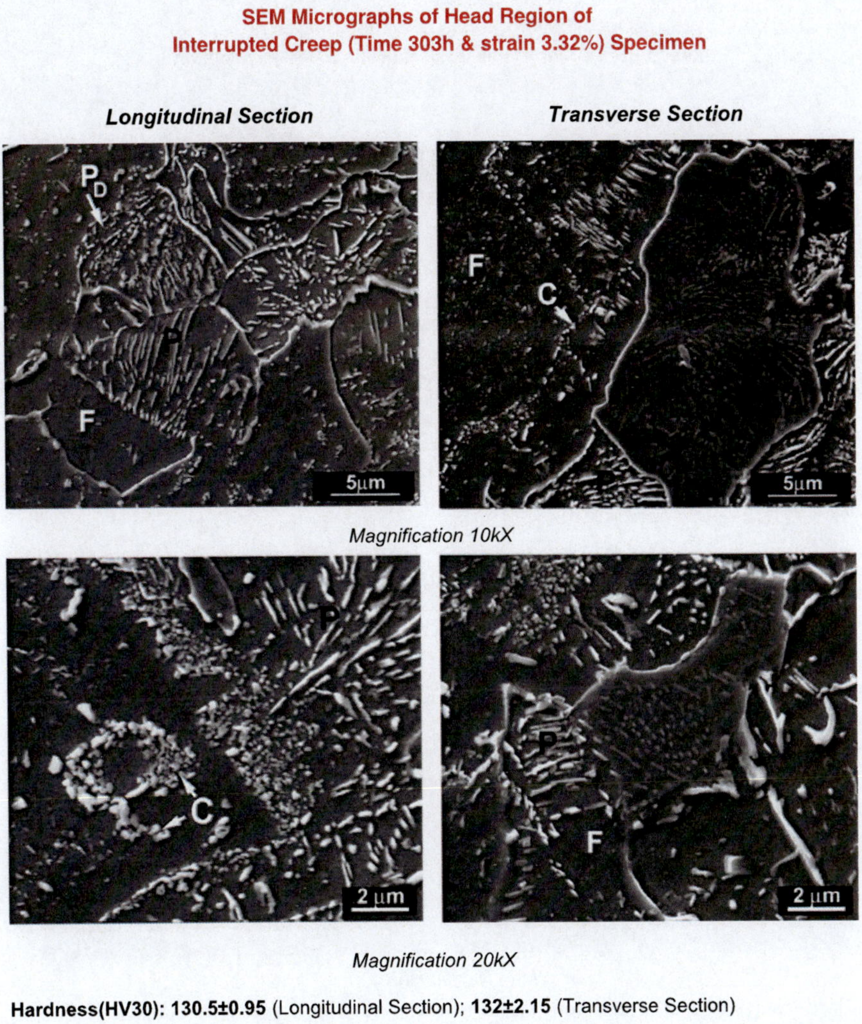

Fig. 4.8 SEM-SE micrographs of **longitudinal and transverse sections** of P22 (2.25Cr–1Mo) steel after creep exposed at 550 °C/115 MPa for 303 h, showing pearlite (P) colonies, discrete pearlite (P_D) and (Fe, Cr, Mo)-rich carbides (C) in ferrite (F) matrix

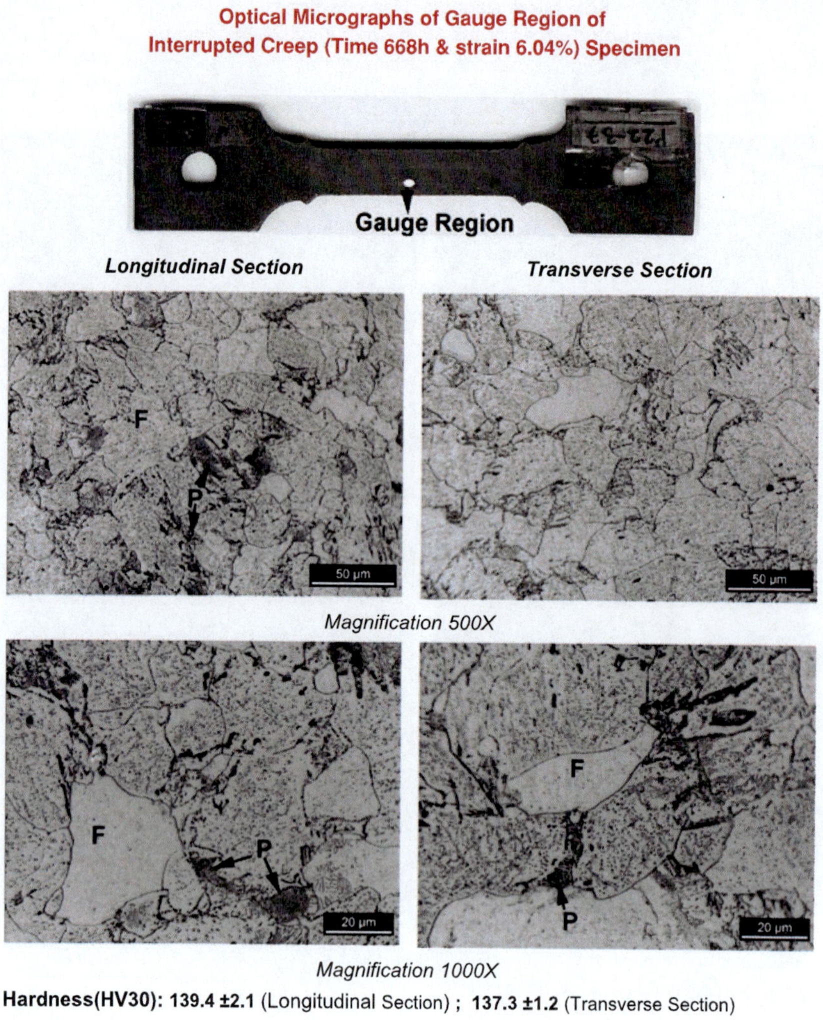

Fig. 4.9 Optical micrographs of **longitudinal and transverse sections** of P22 (2.25Cr–1Mo) steel after creep exposed at 550 °C/115 MPa for 668 h, showing pearlite (P) distributed inhomogeneously in ferrite (F) matrix

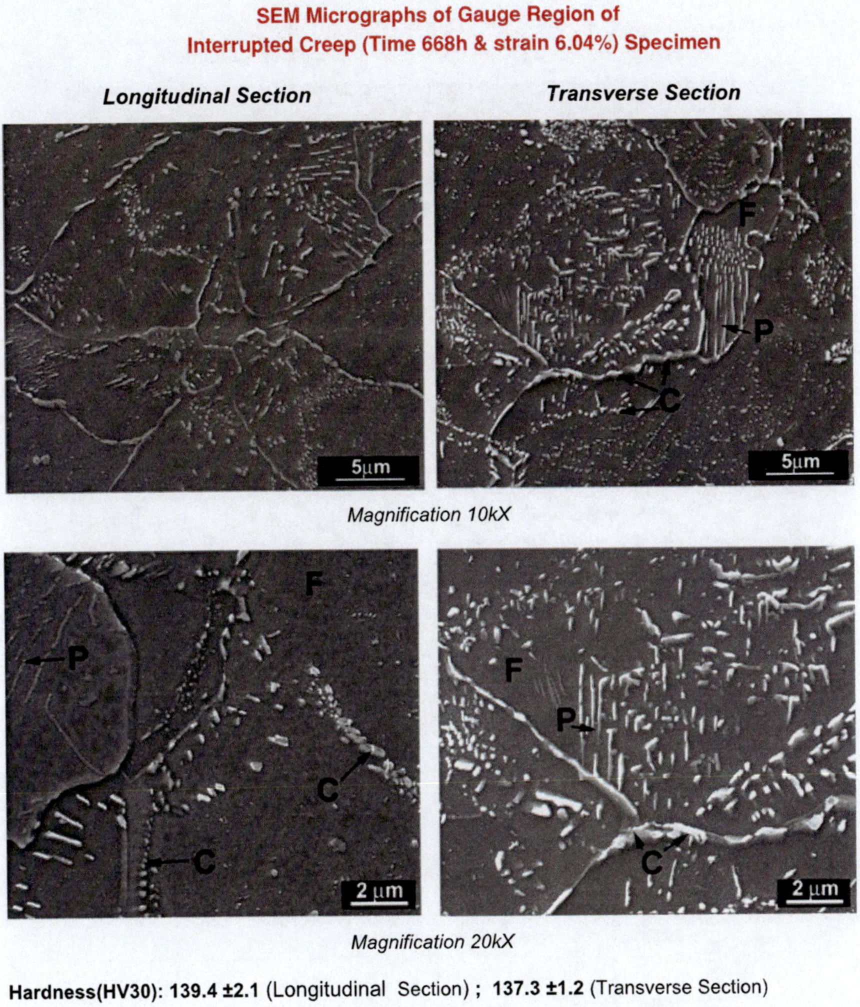

Fig. 4.10 SEM-SE micrographs of **longitudinal and transverse sections** of P22 (2.25Cr–1Mo) steel after creep exposed at 550 °C/115 MPa for 668 h, showing carbide (C) precipitates at grain interiors and grain boundaries, and inhomogeneous distribution of pearlite colonies in ferrite (F) matrix

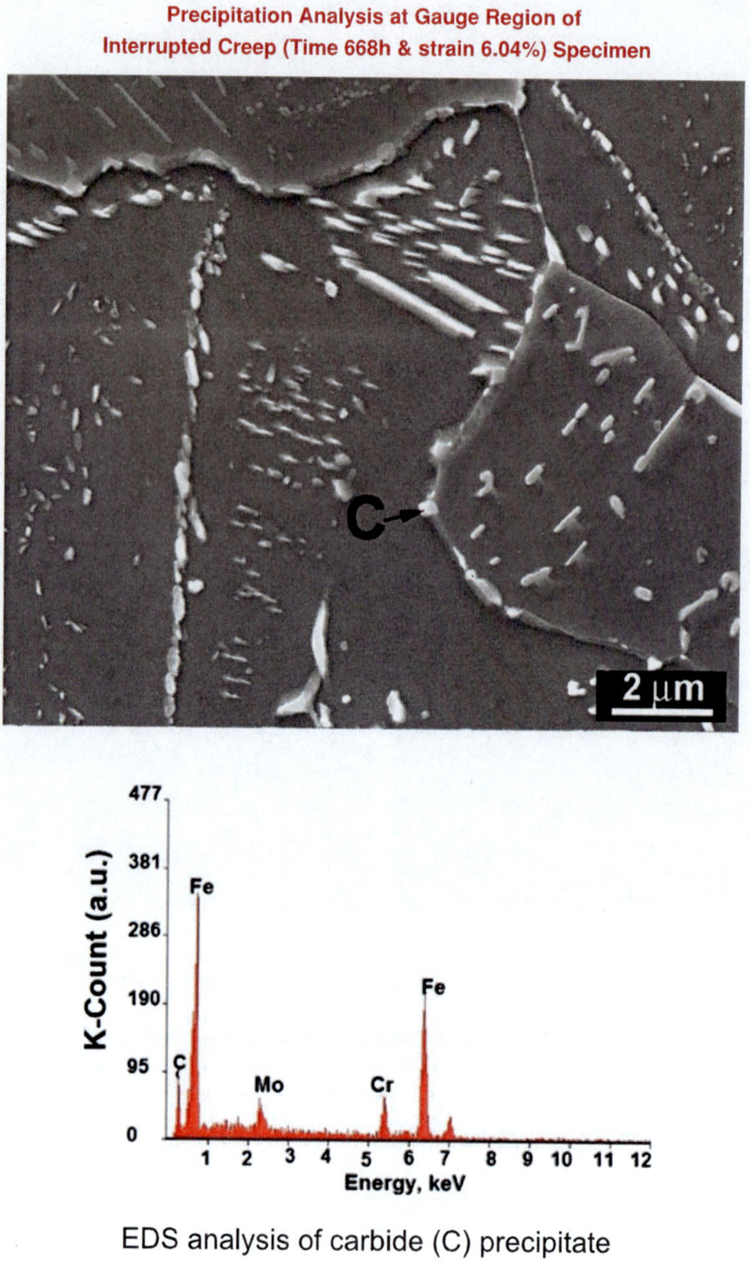

Fig. 4.11 SEM-SE micrograph (magnification 20 kX) of ***transverse sections*** of P22 (2.25Cr–1Mo) steel after creep exposed at 550 °C/115 MPa for 668 h, showing (Fe, Cr, Mo)-rich carbides (C) at grain interior and grain boundaries of ferrite

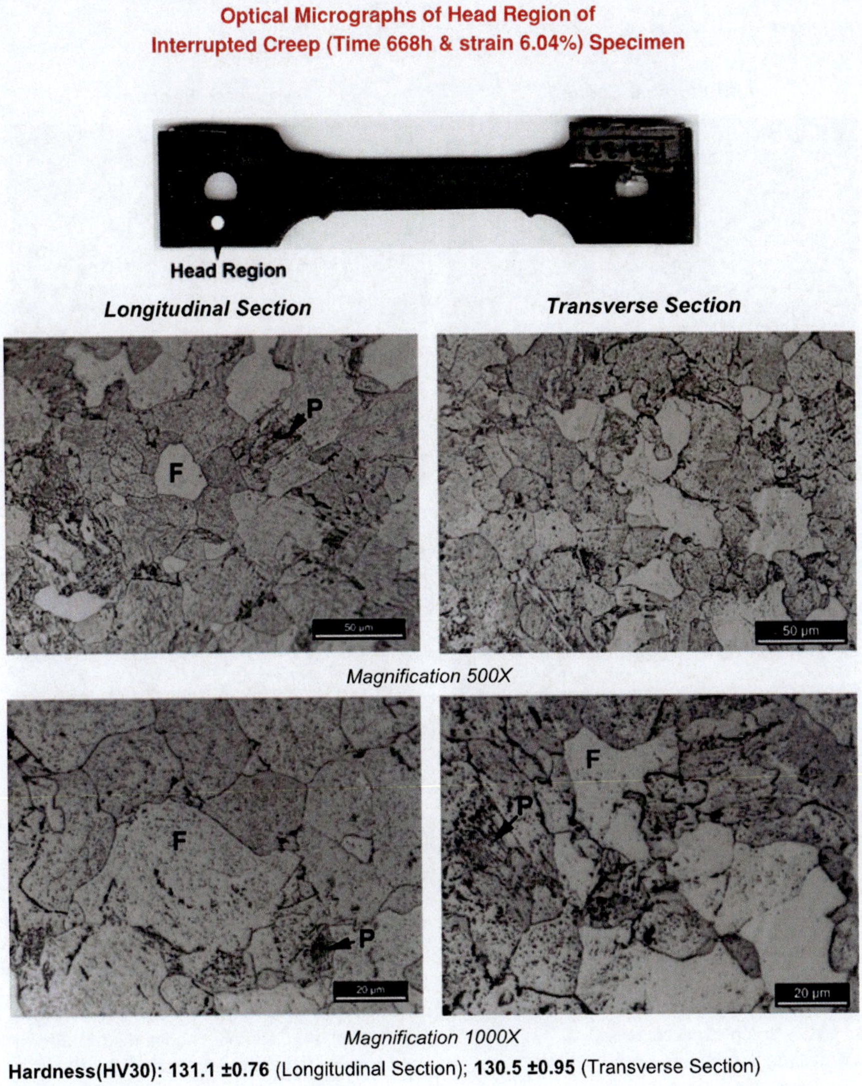

Fig. 4.12 Optical micrographs of **longitudinal and transverse sections** of P22 (2.25Cr–1Mo) steel after creep exposed at 550 °C/115 MPa for 668 h, showing inhomogeneous distribution of ferrite (F) and pearlite (P) structures

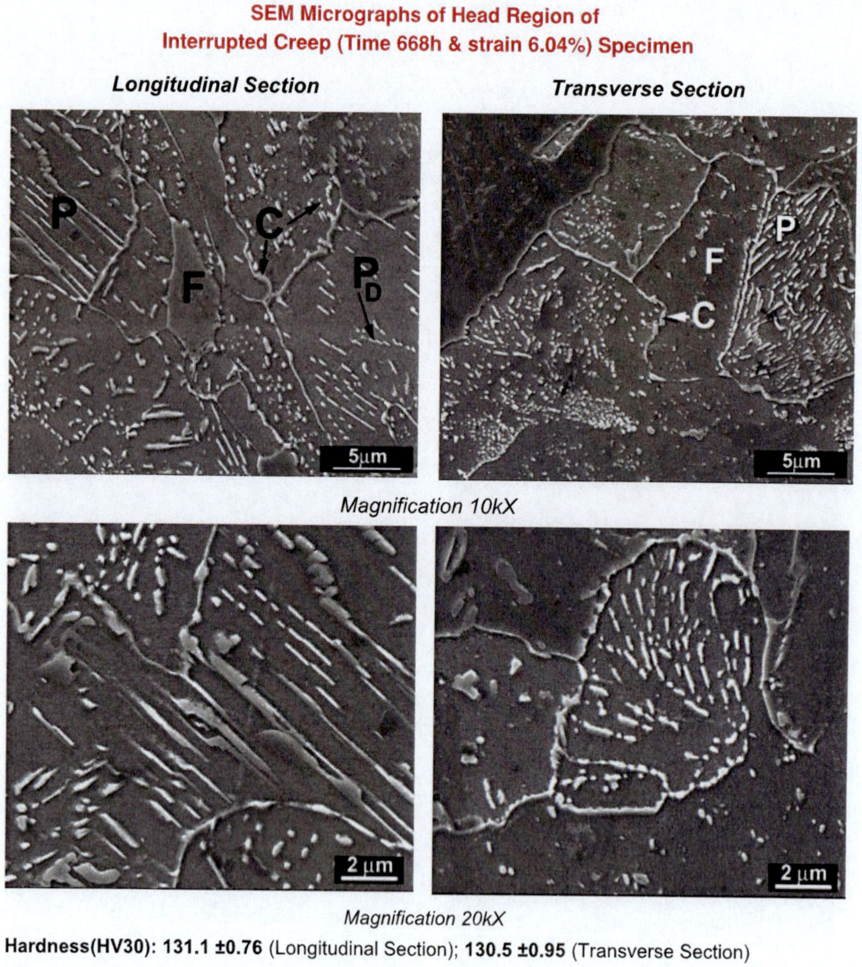

Fig. 4.13 SEM-SE micrographs of **longitudinal and transverse sections** of P22 (2.25Cr–1Mo) steel after creep exposed at 550 °C/115 MPa for 668 h, showing inhomogeneous distribution of carbides (C) at grain interiors and grain boundaries and pearlite (P) colonies, discrete pearlite (P_D) in ferrite matrix

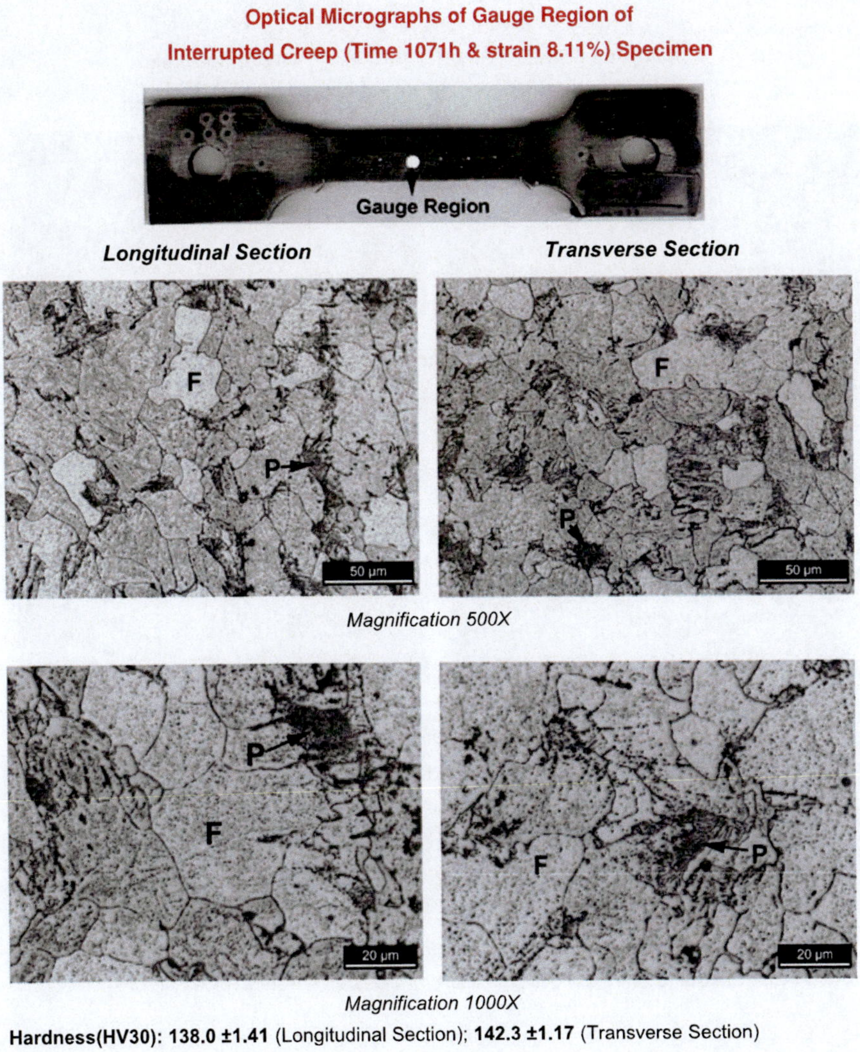

Fig. 4.14 Optical micrographs of **longitudinal and transverse sections** of P22 (2.25Cr–1Mo) steel after creep exposed at 550 °C/115 MPa for 1071 h, showing inhomogeneous distribution of ferrite (F) and pearlite (P) structures

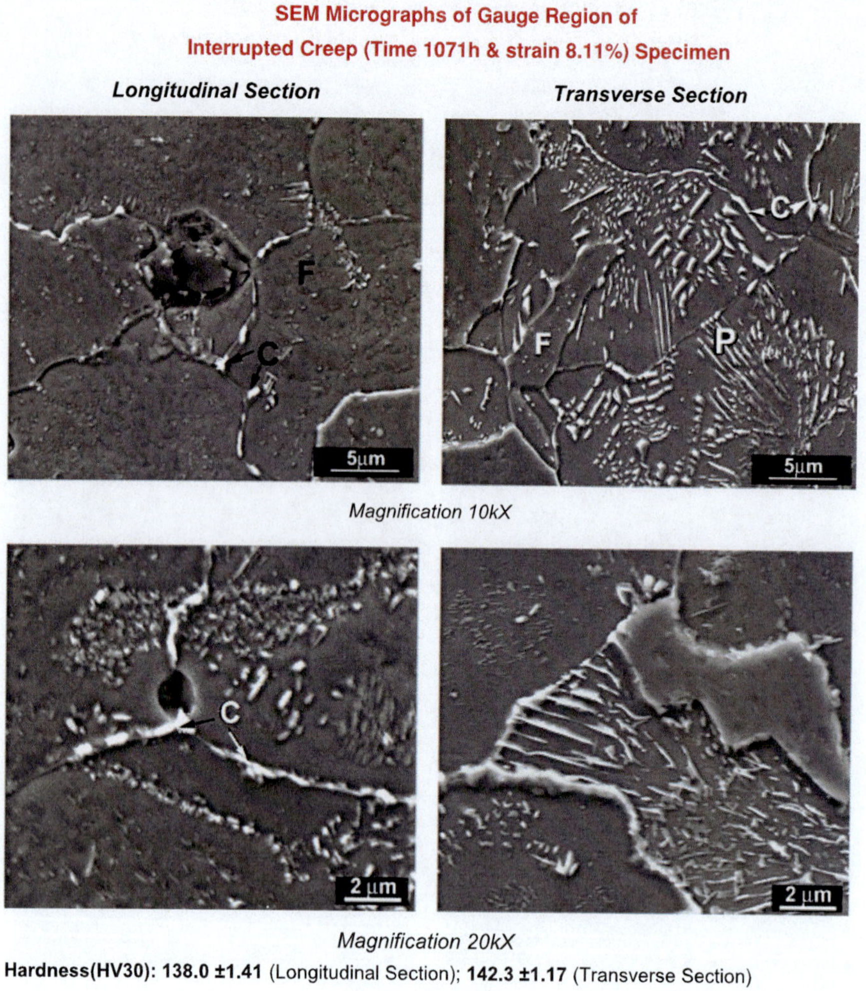

Fig. 4.15 SEM-SE micrographs of **longitudinal and transverse sections** of P22 (2.25Cr–1Mo) steel after creep exposed at 550 °C/115 MPa for 1071 h, explaining carbides (C) at grain interiors and grain boundaries, and pearlite (P) colonies in ferrite matrix

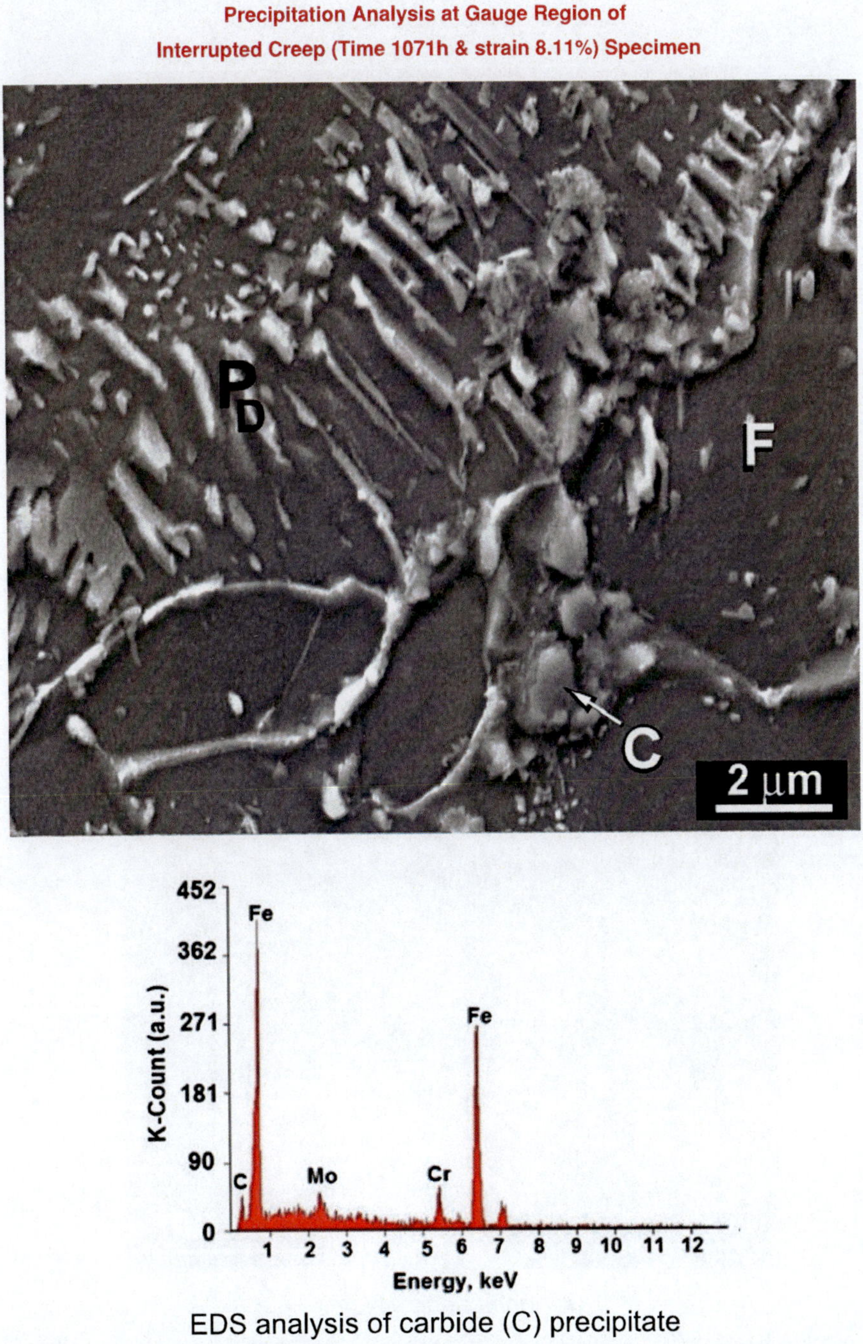

Fig. 4.16 SEM-SE micrograph (magnification 20 kX) of **transverse sections** of P22 (2.25Cr–1Mo) steel after creep exposed at 550 °C/115 MPa for 1071 h, showing (Fe, Cr, Mo)-rich carbide (C) precipitates and discrete pearlite (P_D) distributed at ferrite matrix

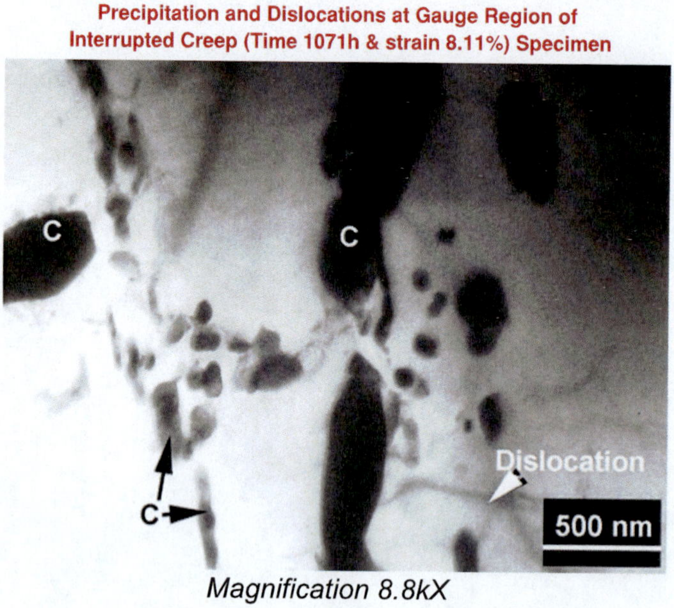

Fig. 4.17 TEM-BF image of <u>*gauge region*</u> at ***transverse section*** of P22(2.25Cr–1Mo) steel after creep exposed at 550 °C/115 MPa for 1071 h with a creep strain of 8.11%, explaining primary and secondary precipitations of globular and parallelepiped carbides (C) at grain boundary and grain interiors, interacting to dislocations

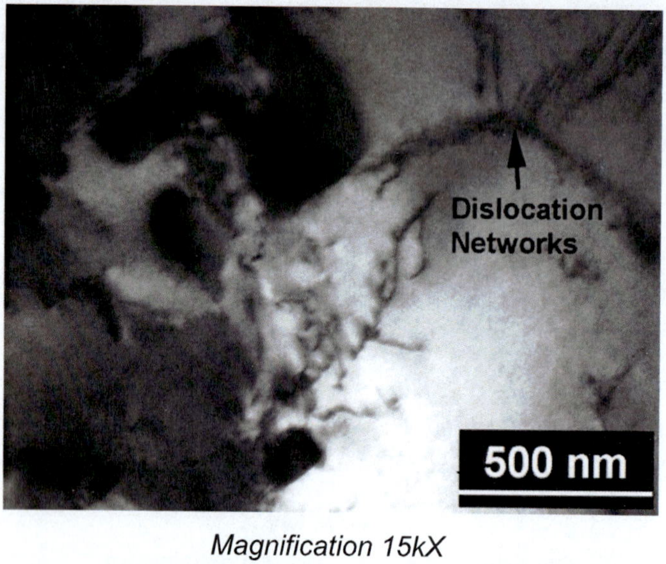

Fig. 4.18 TEM-BF image of <u>*gauge region*</u> at ***transverse section*** of P22(2.25Cr–1Mo) steel after creep exposed at 550 °C/115 MPa for 1071 h with a creep strain of 8.11%, showing dislocation networks in matrix and accumulation of dislocation around large precipitates

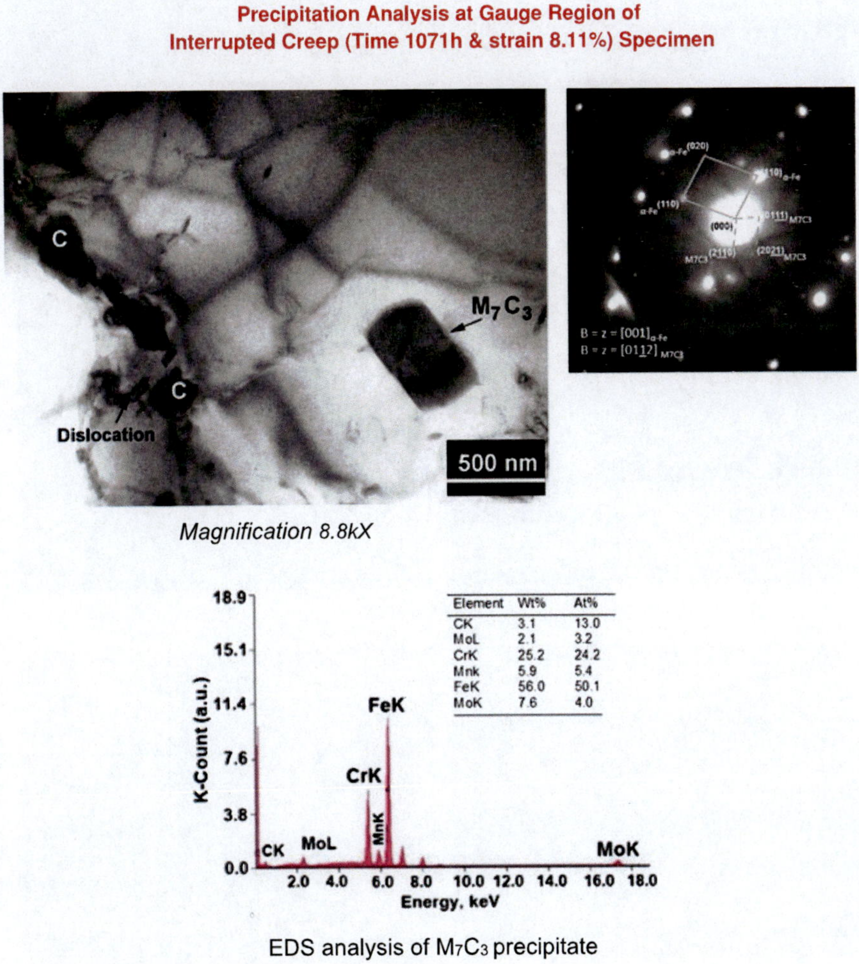

Element	Wt%	At%
CK	3.1	13.0
MoL	2.1	3.2
CrK	25.2	24.2
Mnk	5.9	5.4
FeK	56.0	50.1
MoK	7.6	4.0

Fig. 4.19 TEM-BF image of gauge region at ***transverse section*** of P22(2.25Cr–1Mo) steel after creep exposed at 550 °C/115 MPa for 1071 h with a creep strain of 8.11%, explaining parallelepiped M_7C_3 (with SAED pattern and EDS analysis) and other irregular-shaped carbides (C) precipitated at grain interiors

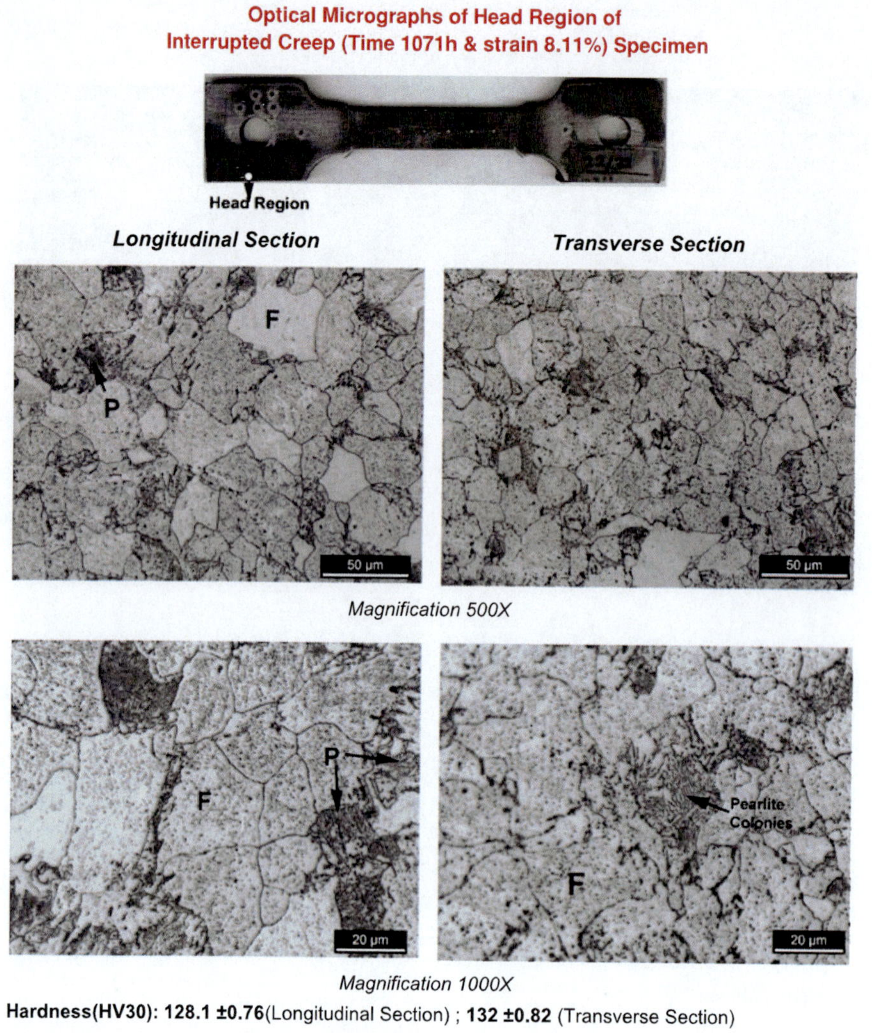

Fig. 4.20 Optical micrographs of **longitudinal and transverse sections** of P22 (2.25Cr–1Mo) steel after creep exposed at 550 °C/115 MPa for 1071 h, showing pearlite (P) colonies in ferrite (F) matrix

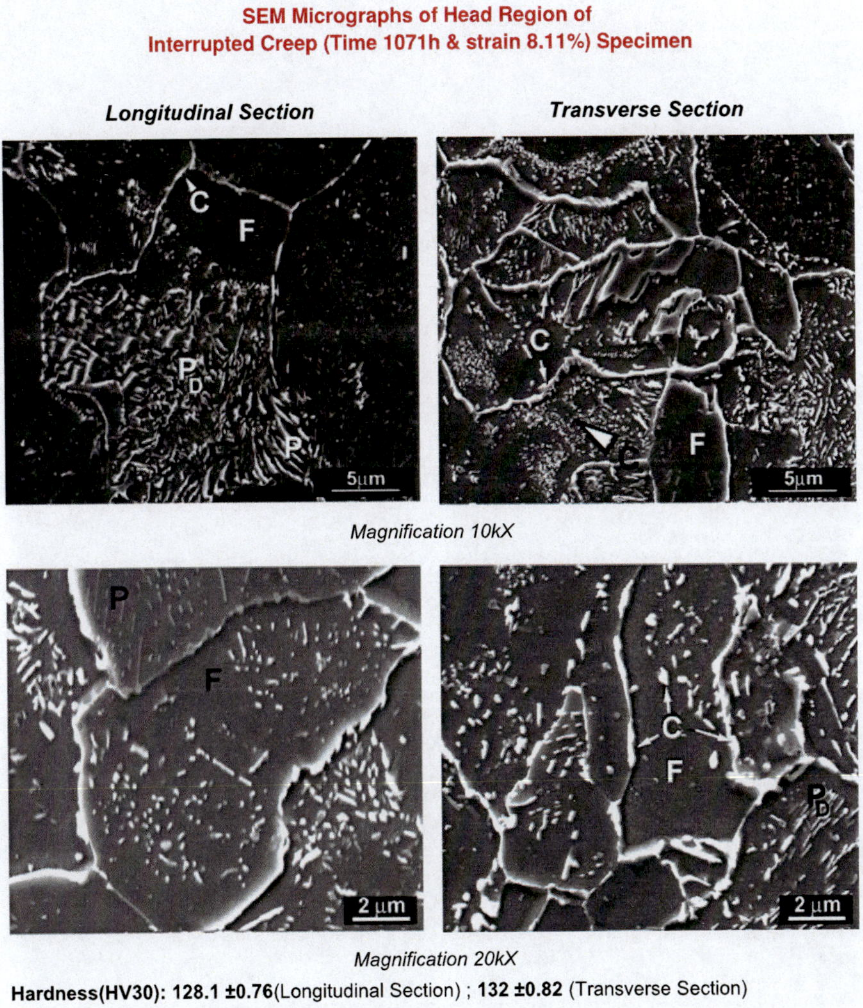

Fig. 4.21 SEM-SE micrographs of **longitudinal and transverse sections** of P22 (2.25Cr–1Mo) steel after creep exposed at 550 °C/115 MPa for 1071 h, in which coarse carbides (C) (black arrows) at grain boundary and fine carbides (white arrow) and discrete pearlite (P_D) at grain interiors, distributed inhomogeneously in ferrite (F) matrix

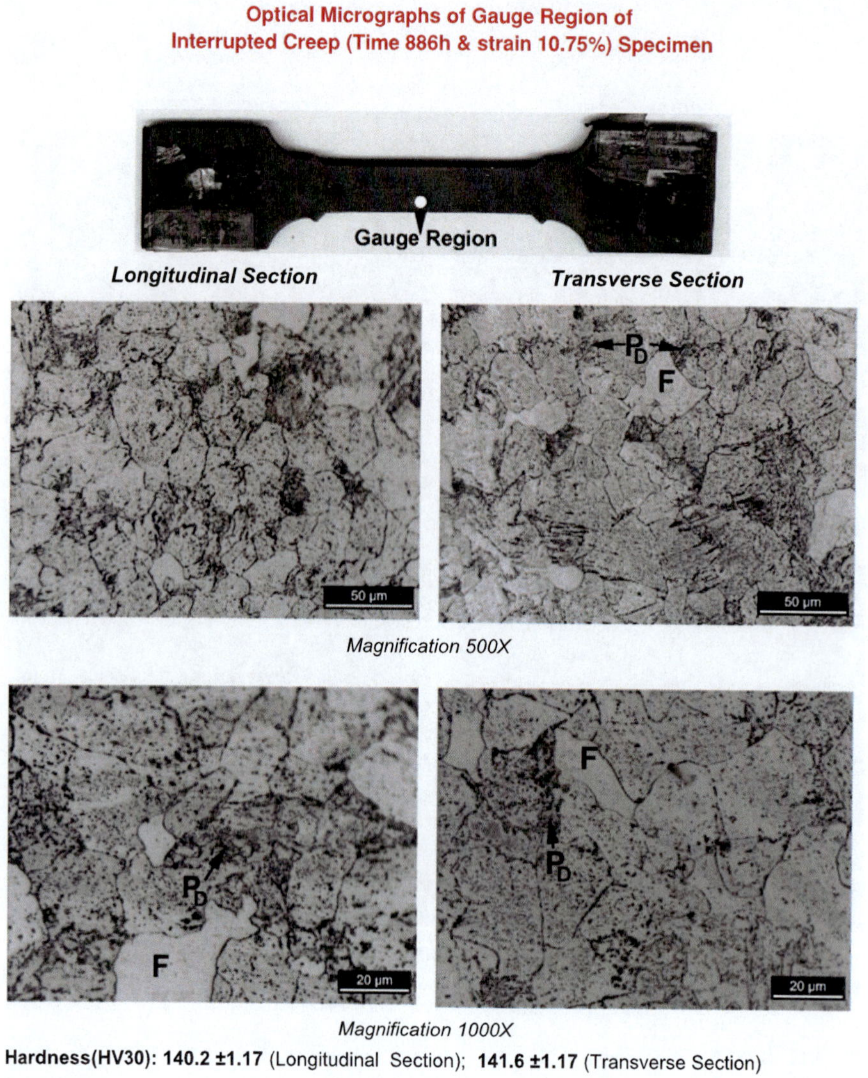

Fig. 4.22 Optical micrographs of **longitudinal and transverse sections** of P22 (2.25Cr–1Mo) steel after creep exposed at 550 °C/115 MPa for 886 h, showing inhomogeneous distribution discrete pearlite (P_D) in ferrite (F) matrix

SEM Micrographs of Gauge Region of Interrupted Creep (Time 886h & strain 10.75%) Specimen

Longitudinal Section ***Transverse Section***

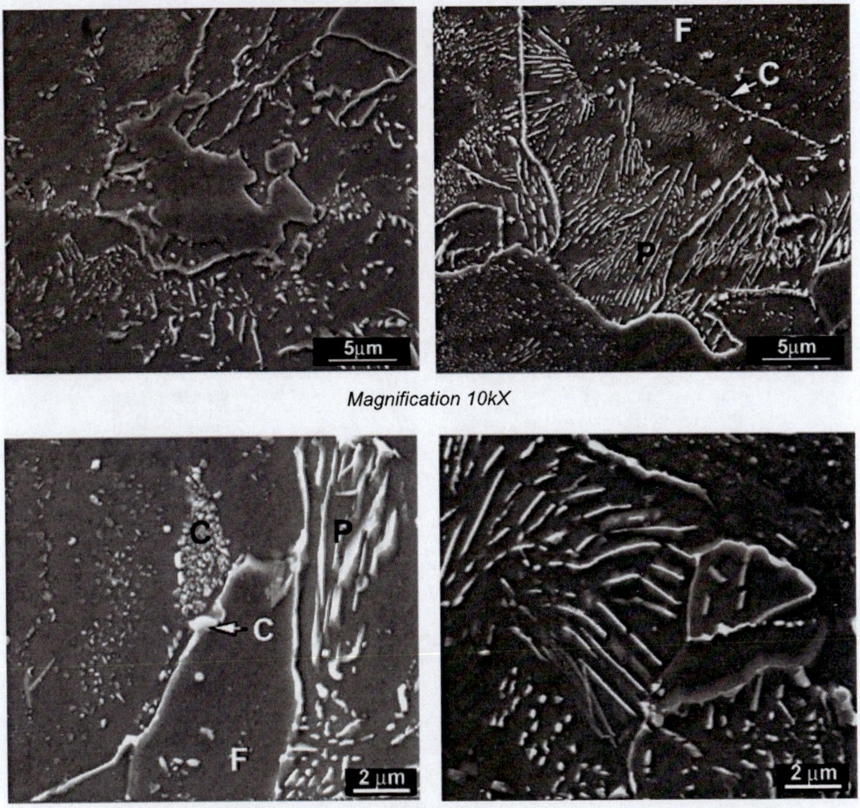

Magnification 20kX

Hardness(HV30): 140.2 ±1.17 (Longitudinal Section); **141.6 ±1.17** (Transverse Section)

Fig. 4.23 SEM-SE micrographs of **longitudinal and transverse sections** of P22 (2.25Cr–1Mo) steel after creep exposed at 550 °C/115 MPa for 886 h, explaining inhomogeneous distribution of coarse carbides (C) at grain boundaries, fine carbides at grain interiors and pearlite (P) colonies in ferrite (F) matrix

Precipitation and Dislocations at Gauge Region of Interrupted Creep (Time 886h & strain 10.75%) Specimen

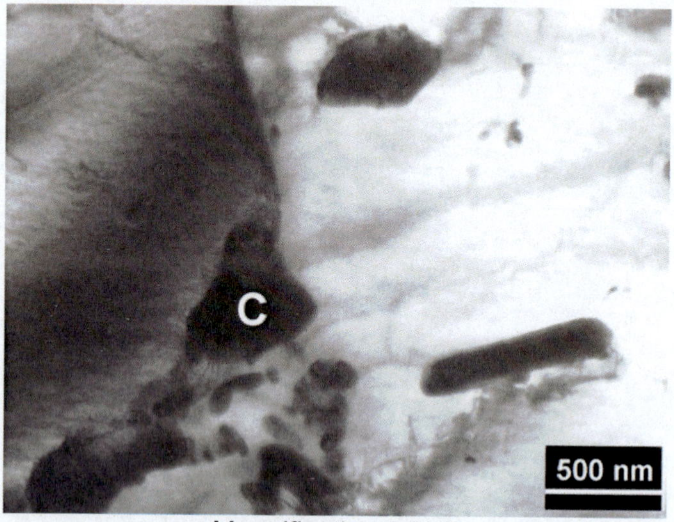

Magnification 8.8kX

Fig. 4.24 TEM-BF image of gauge region at ***transverse section*** of P22(2.25Cr–1Mo) steel after creep exposed at 550 °C/115 MPa for 886 h with a creep strain of 10.75%, explaining angular large (>400 nm) precipitate (C) due to coagulation and secondary precipitation showing near globular/elliptical precipitates

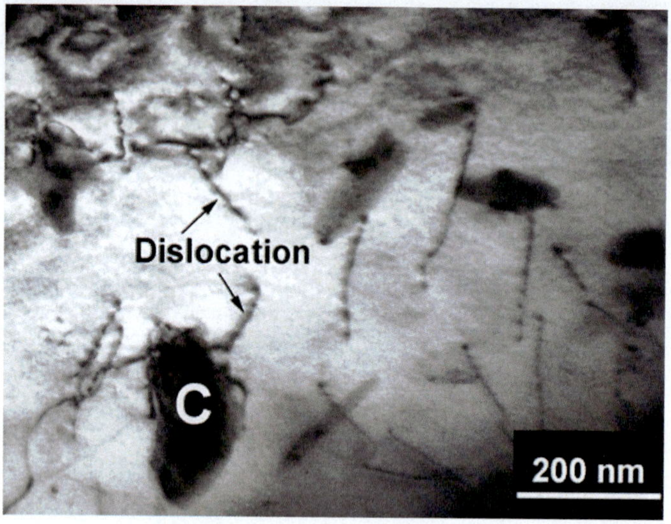

Magnification 27.5kX

Fig. 4.25 TEM-BF image of gauge region at ***transverse section*** of P22(2.25Cr–1Mo) steel after creep exposed at 550 °C/115 MPa for 886 h with a creep strain of 10.75%, showing dislocation lines in matrix and pinning of dislocation by carbide precipitate (C)

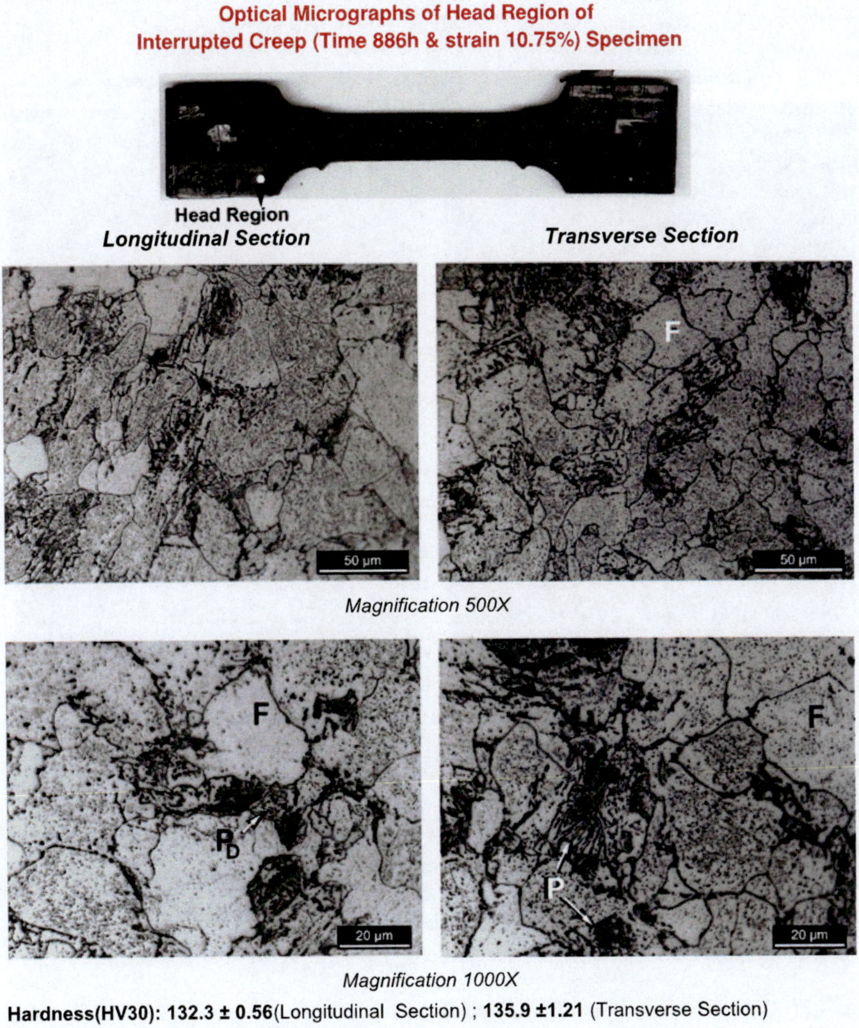

Fig. 4.26 Optical micrographs of **longitudinal and transverse sections** of P22 (2.25Cr–1Mo) steel after creep exposed at 550 °C/115 MPa for 886 h, showing pearlite (P) colonies and discrete pearlite (P_D) in ferrite (F) matrix

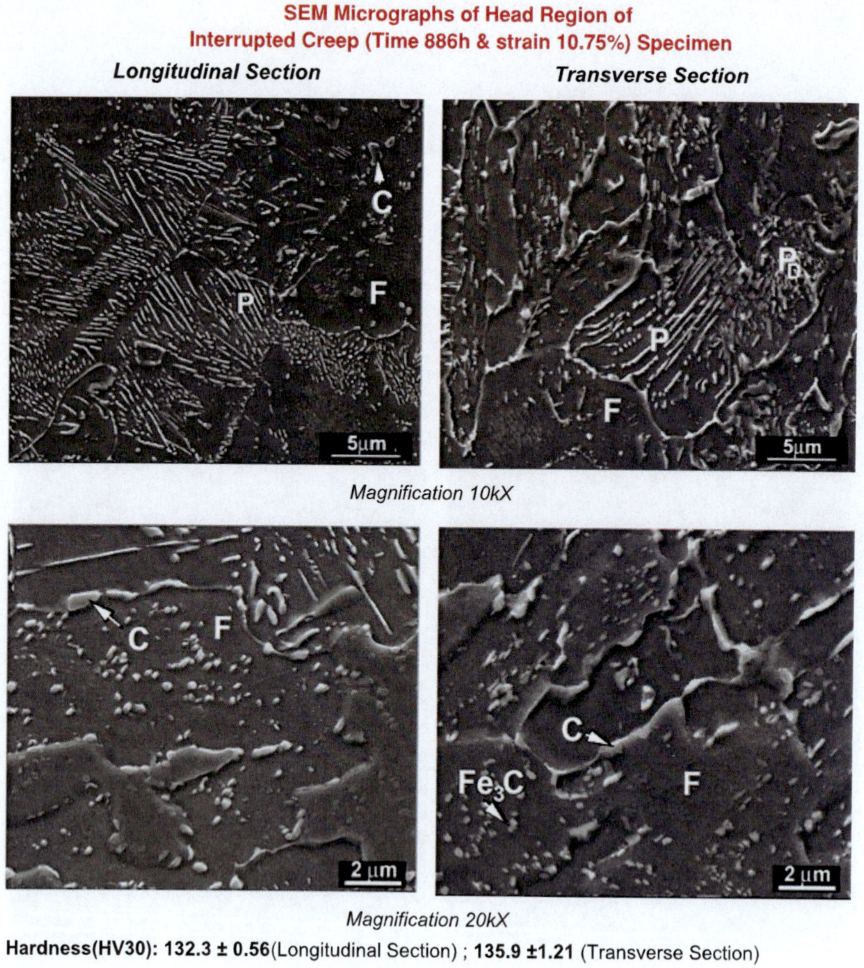

Fig. 4.27 SEM-SE micrographs of **longitudinal and transverse sections** of P22 (2.25Cr–1Mo) steel after creep exposed at 550 °C/115 MPa for 886 h, explaining coarse carbides (C) at grain boundaries, cementite (Fe_3C), pearlite (P) colonies and discrete pearlite (P_D) in ferrite (F) matrix

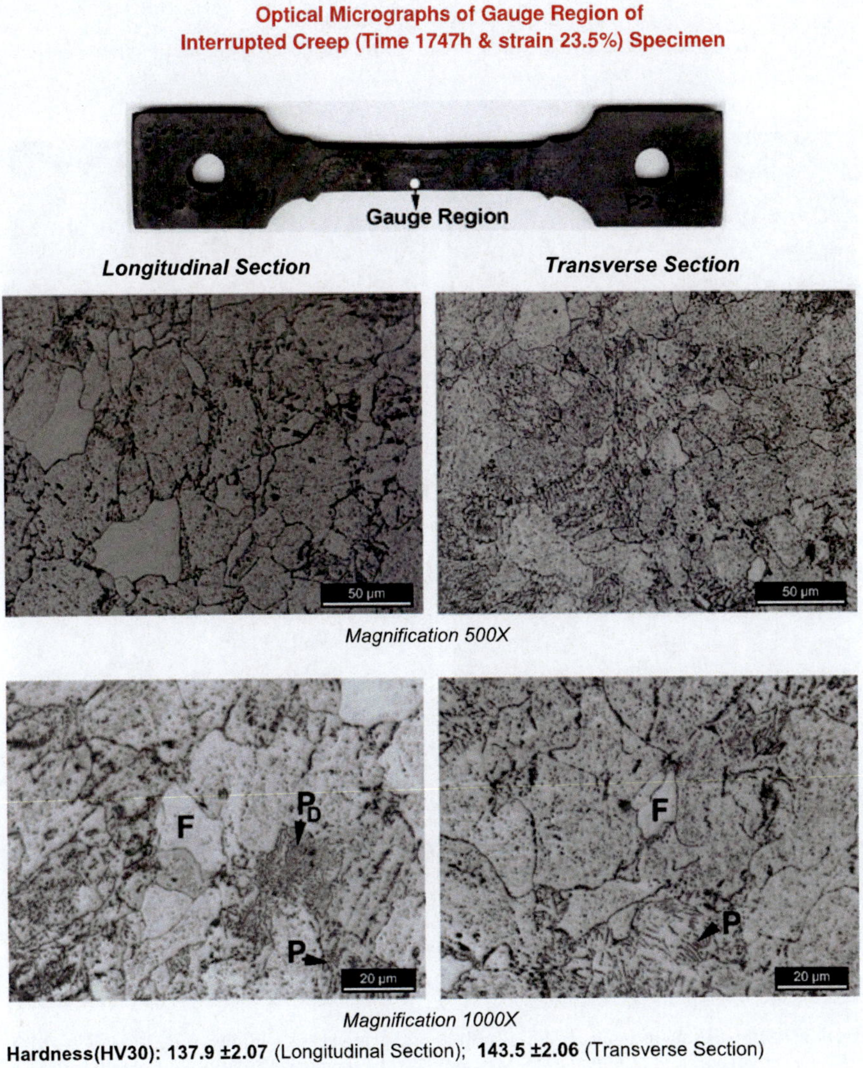

Fig. 4.28 Optical micrographs of **longitudinal and transverse sections** of P22 (2.25Cr–1Mo) steel after creep exposed at 550 °C/115 MPa for 1747 h, showing inhomogeneous distribution of pearlite (P) colonies and discrete pearlite (P_D) in ferrite (F) matrix

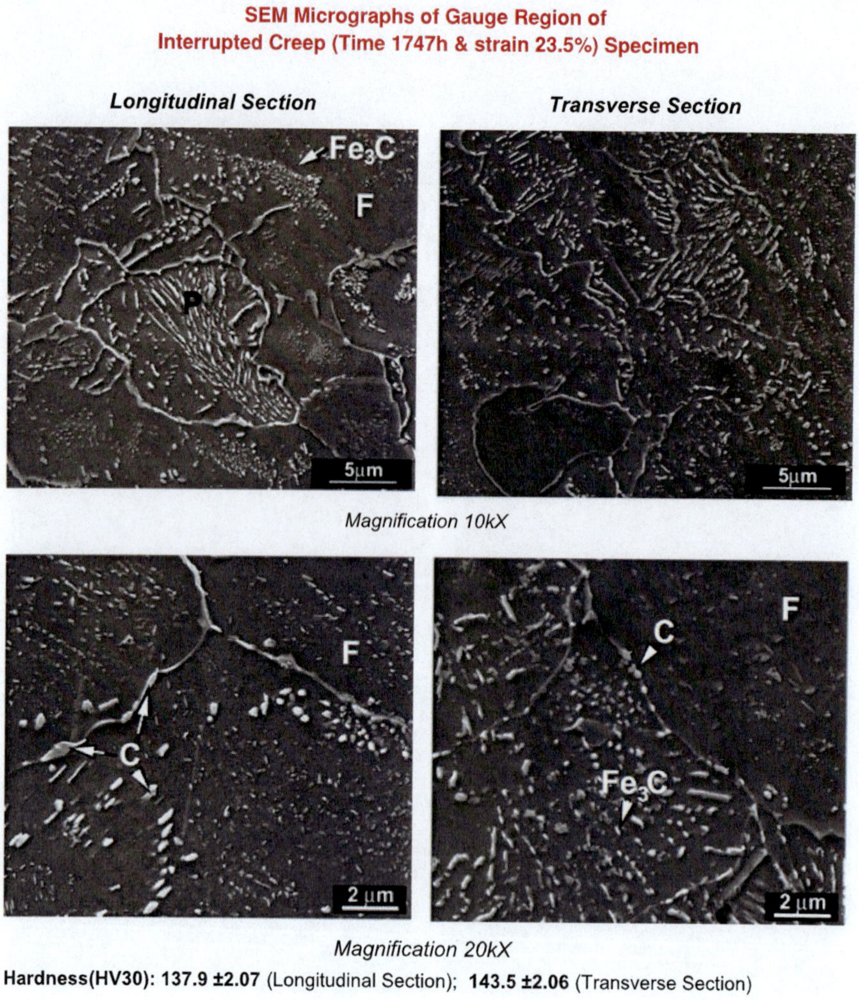

Fig. 4.29 SEM-SE micrographs of **longitudinal and transverse sections** of P22 (2.25Cr–1Mo) steel after creep exposed at 550 °C/115 MPa for 1747 h, showing distribution of pearlite (P) and fine Fe_3C in ferrite (F) matrix and coarse carbides (C) precipitates at grain boundaries

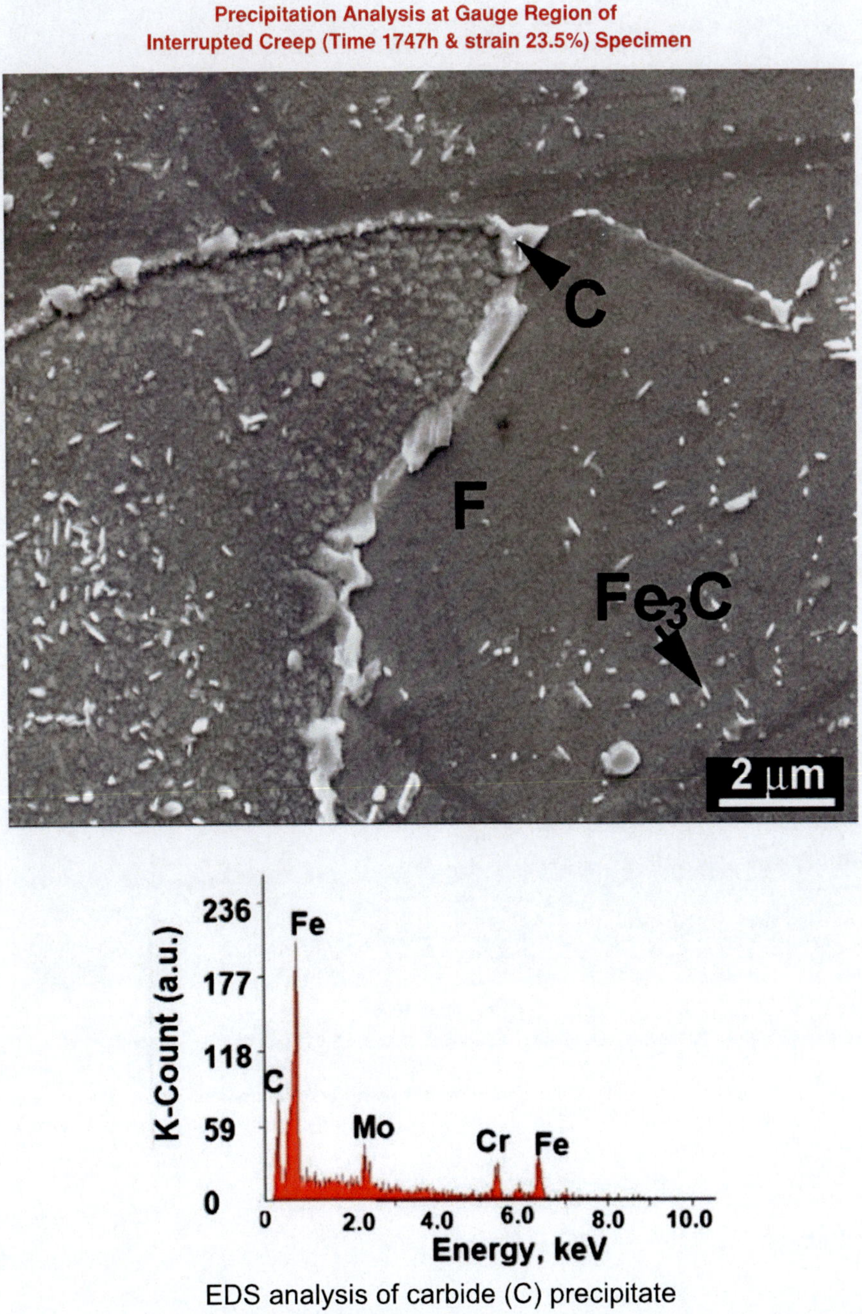

Fig. 4.30 SEM-SE micrograph (magnification 20 kX) of **transverse sections** of P22 (2.25Cr–1Mo) steel after creep exposed at 550 °C/115 MPa for 1747 h, presenting fine Fe_3C in ferrite (F) matrix and coarse (Fe, Cr, Mo)-rich carbides (C) precipitates at grain boundaries

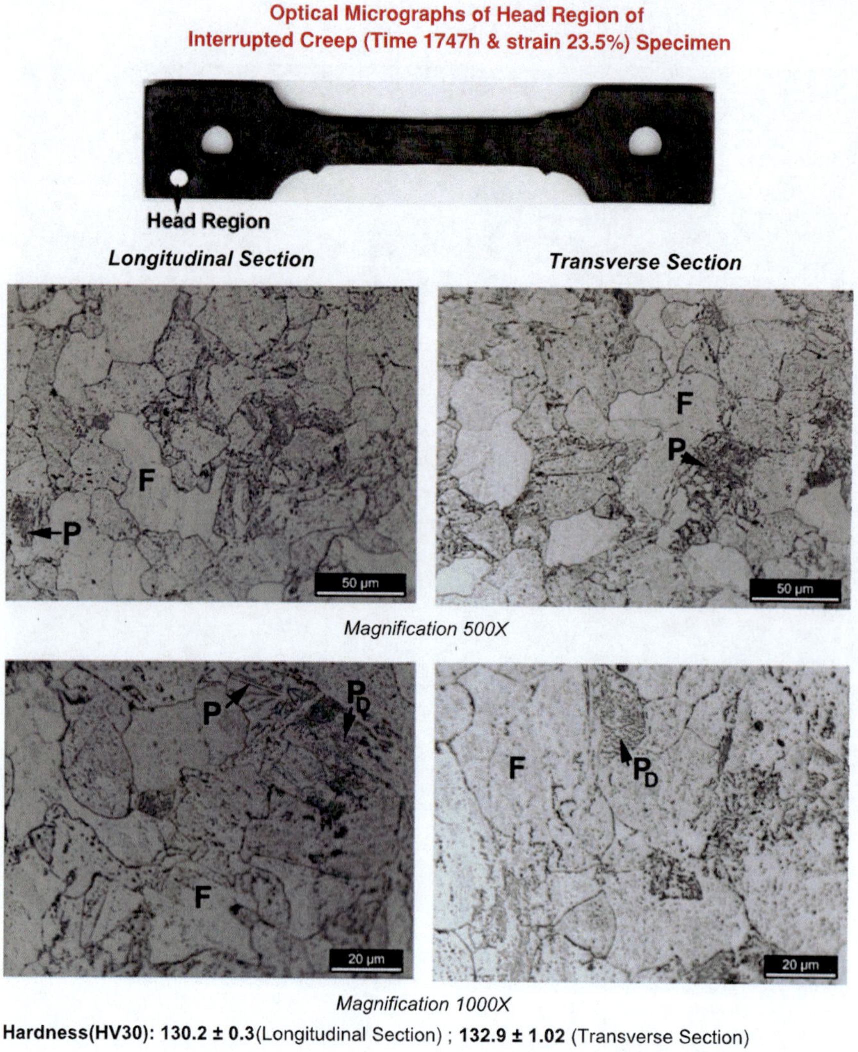

Fig. 4.31 Optical micrographs of **longitudinal and transverse sections** of P22 (2.25Cr–1Mo) steel after creep exposed at 550 °C/115 MPa for 1747 h, showing pearlite (P) colonies and discrete pearlite (P_D) in ferrite (F) matrix

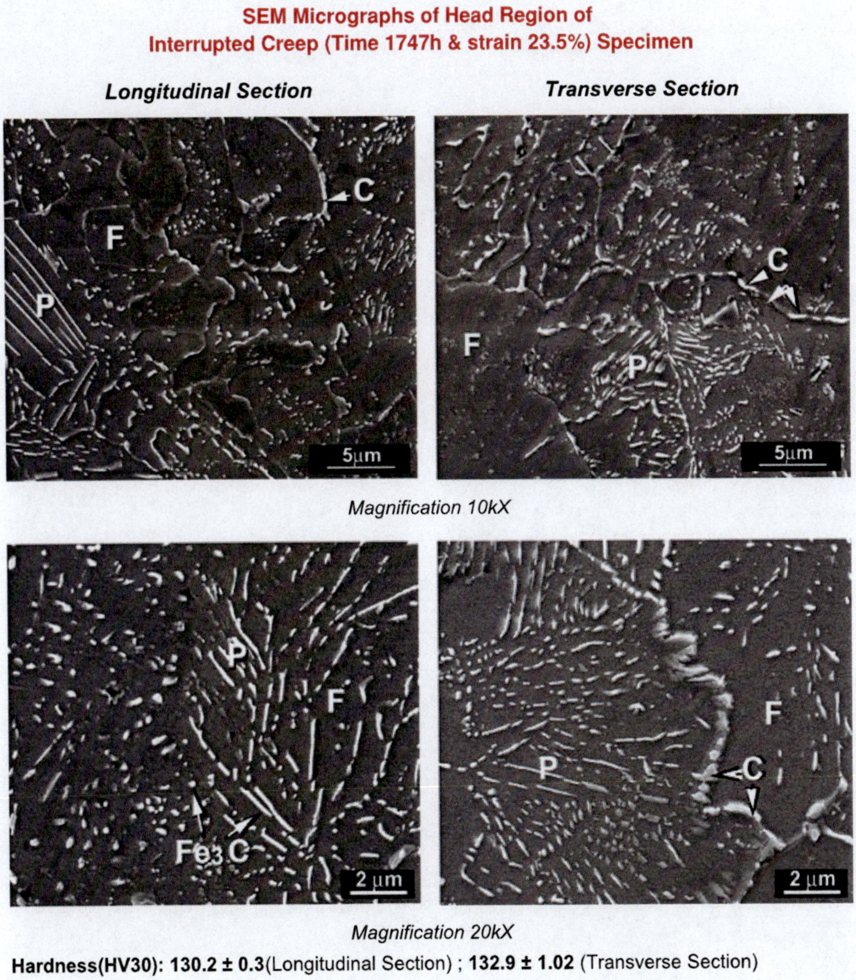

Fig. 4.32 SEM-SE micrographs of **longitudinal and transverse sections** of P22 (2.25Cr–1Mo) steel after creep exposed at 550 °C/115 MPa for 1747 h, showing formation of pearlite (P) phase in which fine cementite (Fe_3C) distributed in ferrite (F) matrix and coarse carbides (C) at grain boundaries

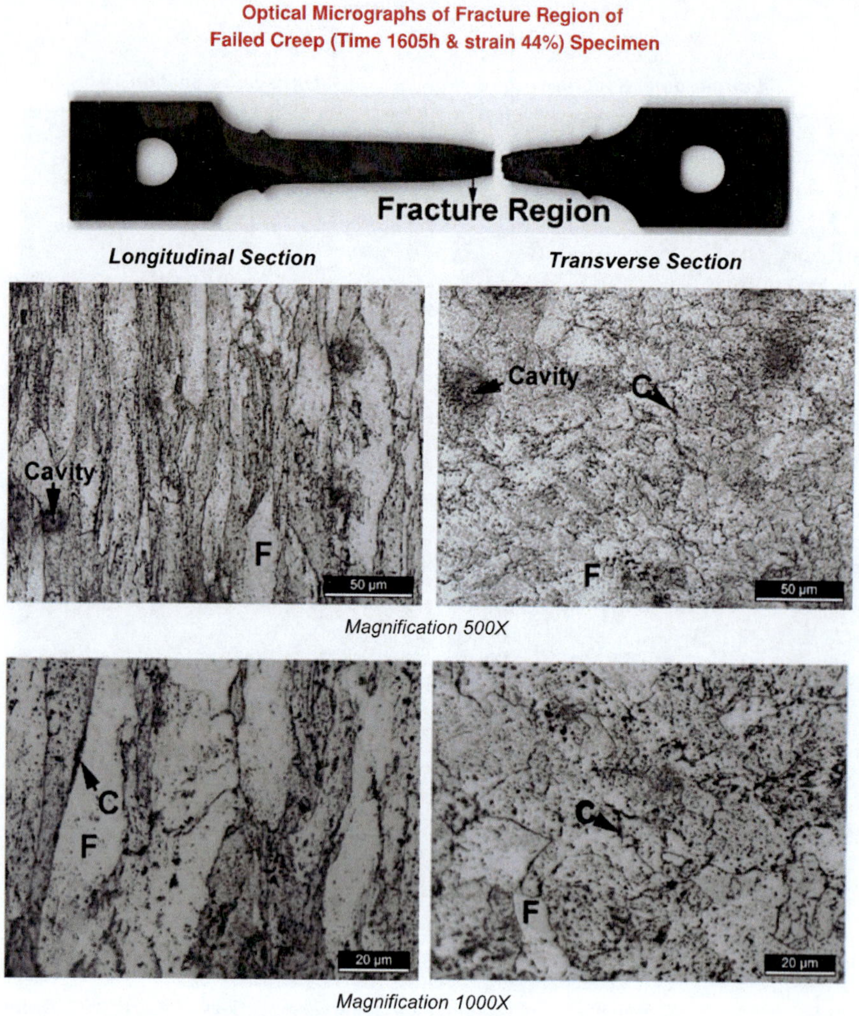

Fig. 4.33 Optical micrographs of **longitudinal and transverse sections** of P22 (2.25Cr–1Mo) steel after *creep failed* at 550 °C/115 MPa for 1605 h, showing cavity formation in ferrite (F) matrix and carbide (C) precipitates at grain boundaries

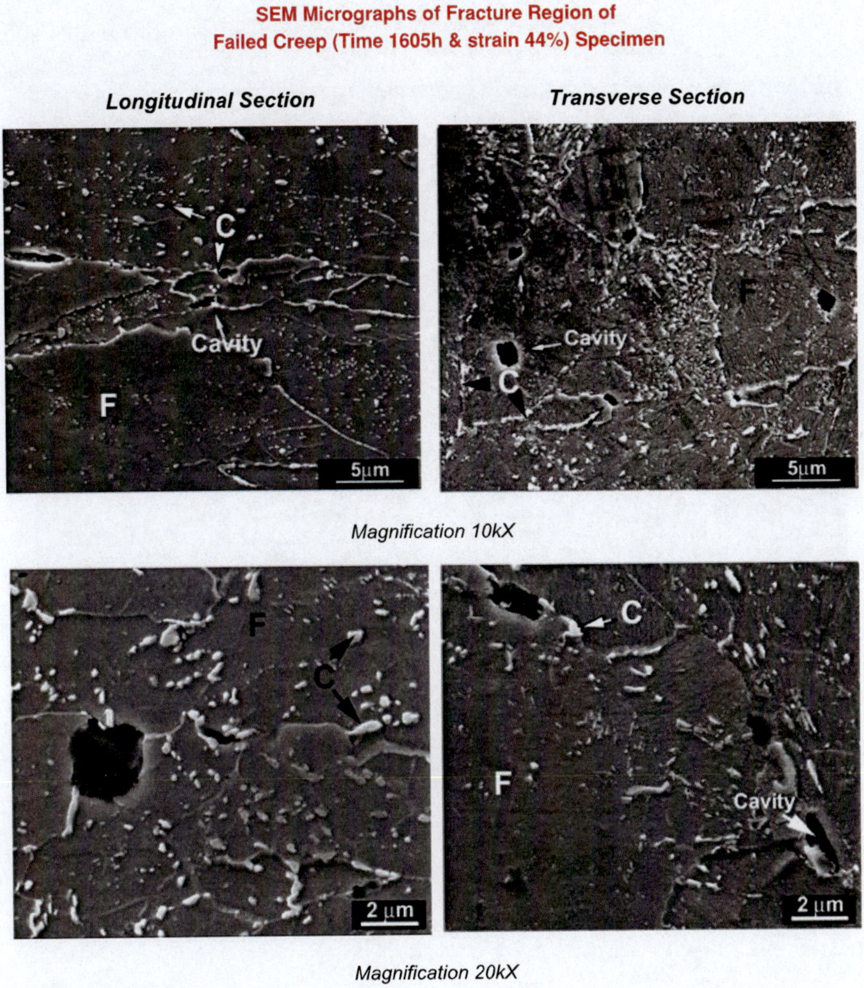

Fig. 4.34 SEM-SE micrographs of **longitudinal and transverse sections** of P22 (2.25Cr–1Mo) steel after *creep failed* at 550 °C/115 MPa for 1605 h, showing inhomogeneous distribution of (Fe, Cr, Mo)-rich carbides (C) and cavity (white arrows) in ferrite (F) matrix

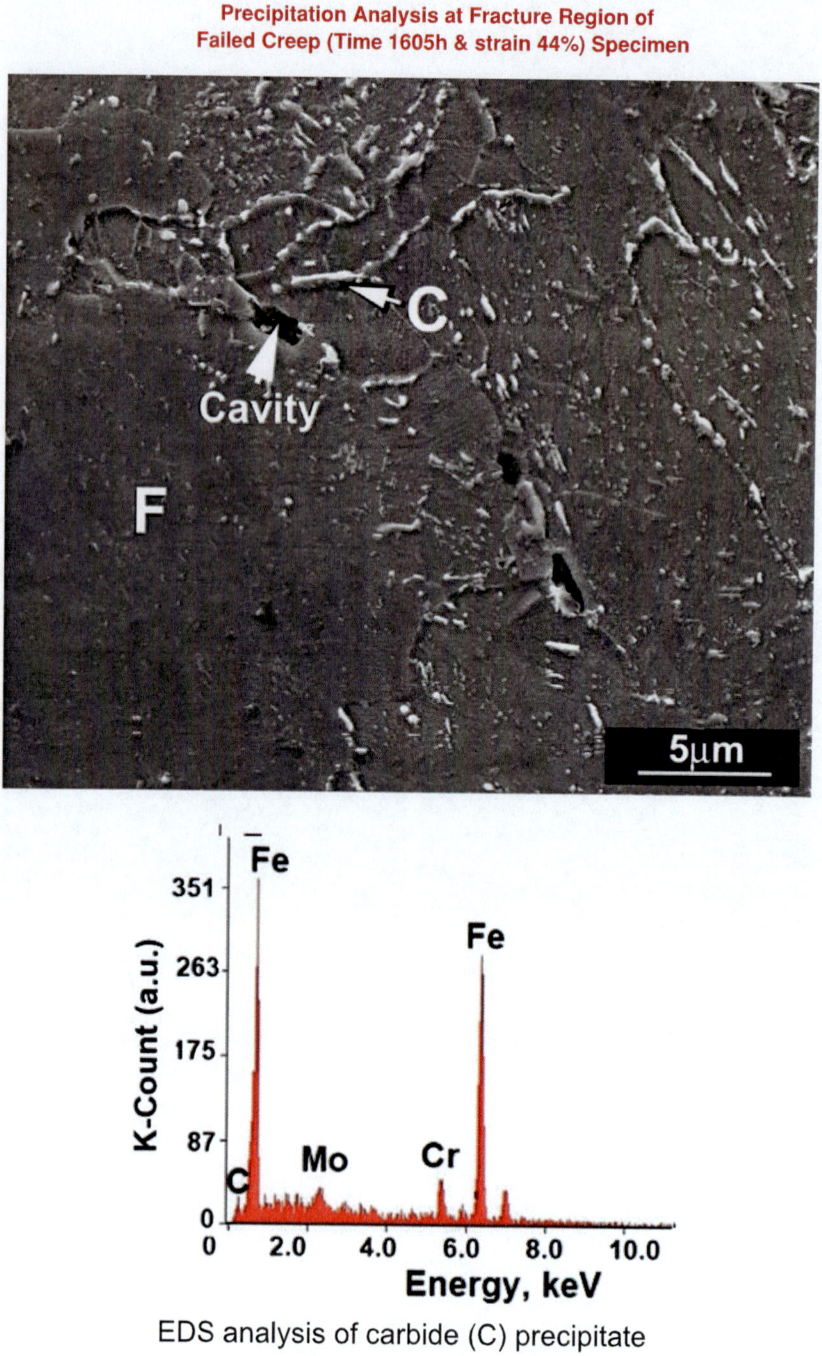

Fig. 4.35 SEM-SE micrograph (magnification 10 kX) of **transverse sections** of P22 (2.25Cr–1Mo) steel after <u>*creep failed*</u> at 550 °C/115 MPa for 1605 h, showing inhomogeneous distribution of (Fe, Cr, Mo)-rich carbides (C) and cavity (white arrows) in ferrite (F) matrix

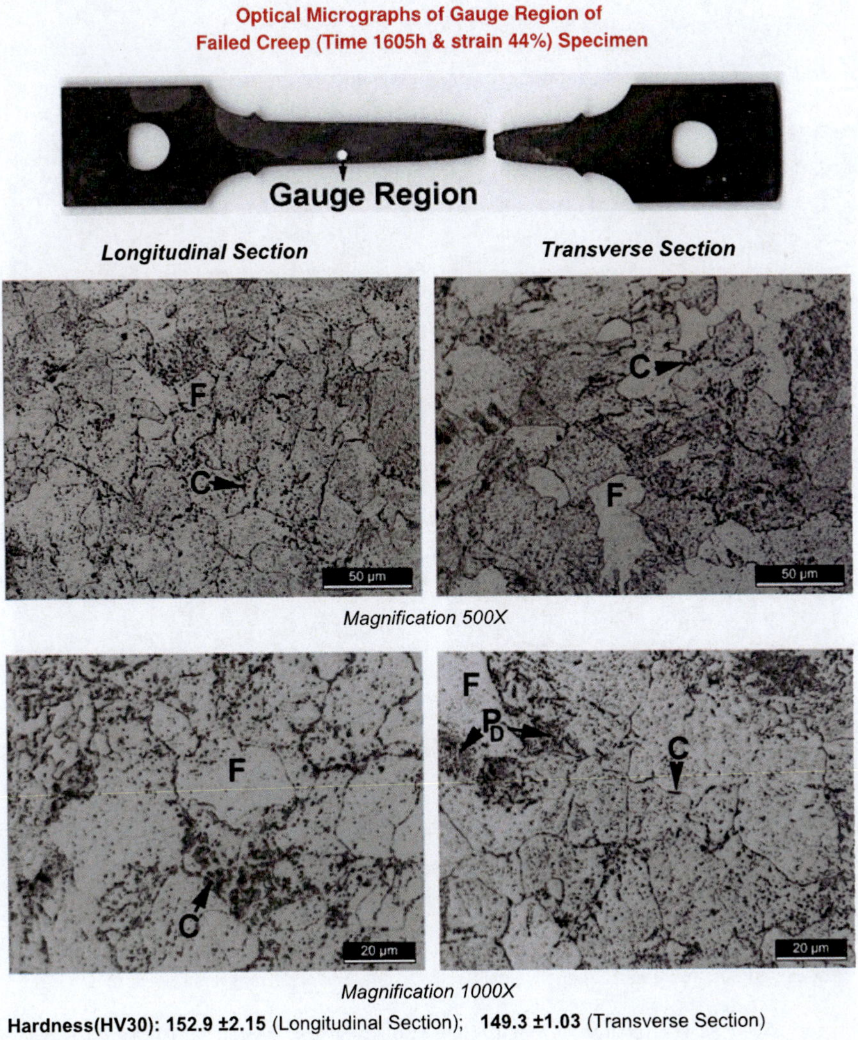

Fig. 4.36 Optical micrographs of **longitudinal and transverse sections** of P22 (2.25Cr–1Mo) steel after creep failed at 550 °C/115 MPa for 1605 h, explaining discrete pearlite (P_D) and carbide (C) precipitates at grain boundaries in ferrite (F) matrix

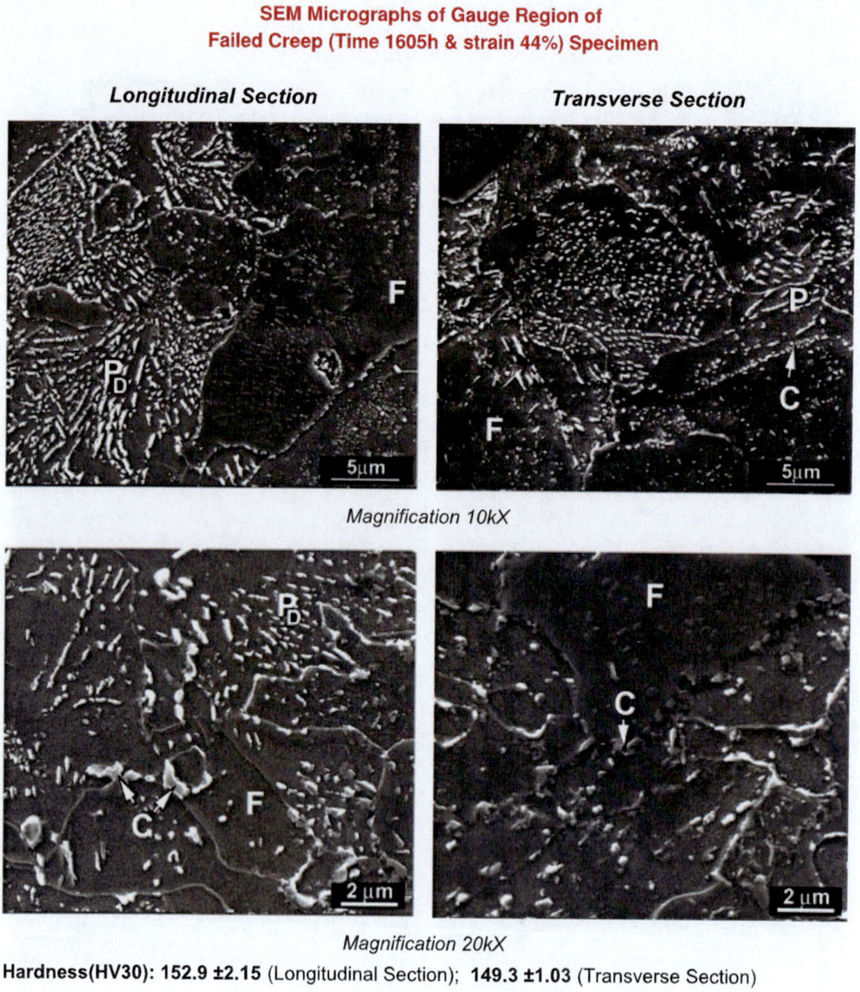

Fig. 4.37 SEM-SE micrographs of **longitudinal and transverse sections** of P22 (2.25Cr–1Mo) steel after creep failed at 550 °C/115 MPa for 1605 h, showing inhomogeneous distribution of (Fe, Cr, Mo)-rich carbides (C) and discrete pearlite (P_D) in ferrite (F) matrix

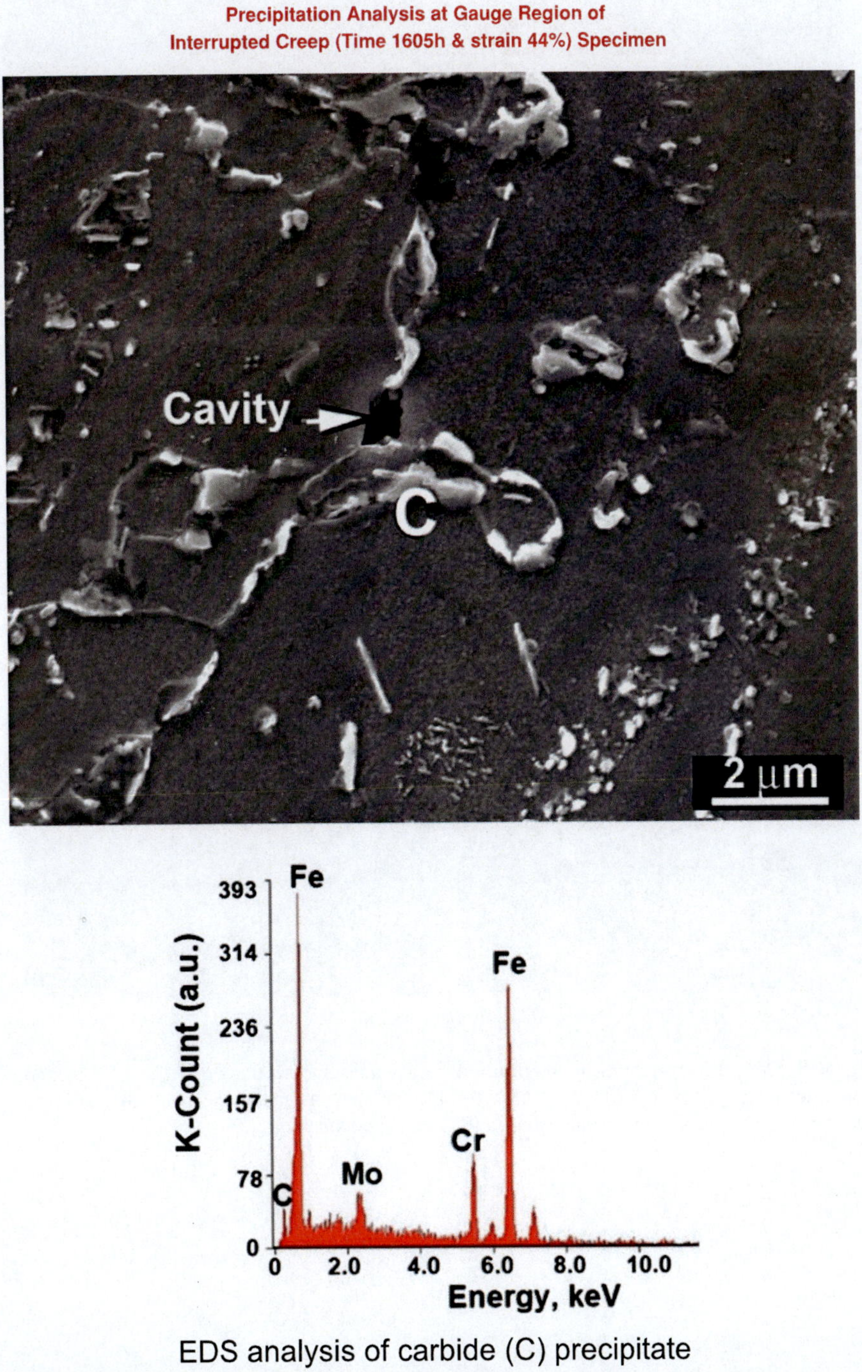

Fig. 4.38 SEM-SE micrograph (magnification 20 kX) of **longitudinal sections** of P22 (2.25Cr–1Mo) steel after *creep failed* at 550 °C/115 MPa for 1605 h, showing the initiation of cavity near (Fe, Cr, Mo)-rich carbides (C) in ferrite (F) matrix

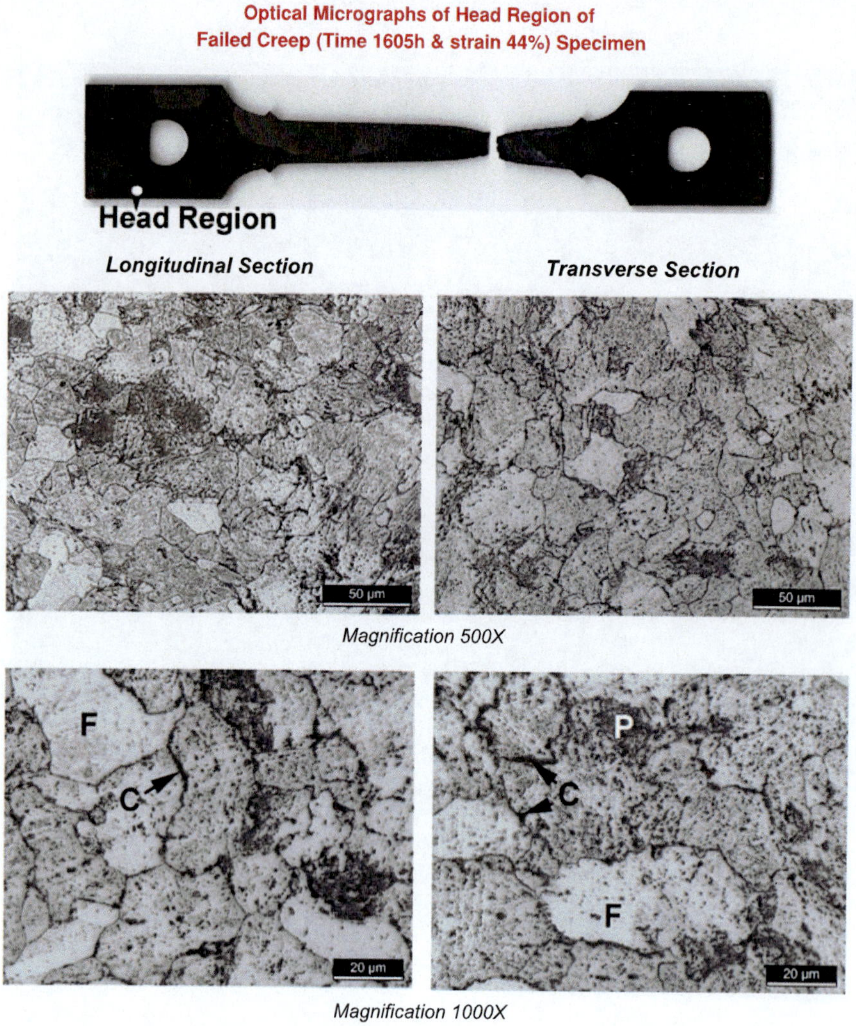

Fig. 4.39 Optical micrographs of **longitudinal and transverse sections** of P22 (2.25Cr–1Mo) steel after *creep failed* at 550 °C/115 MPa for 1605 h, showing inhomogeneous distribution of pearlite (P) in ferrite (F) matrix and carbides (C) at grain boundaries

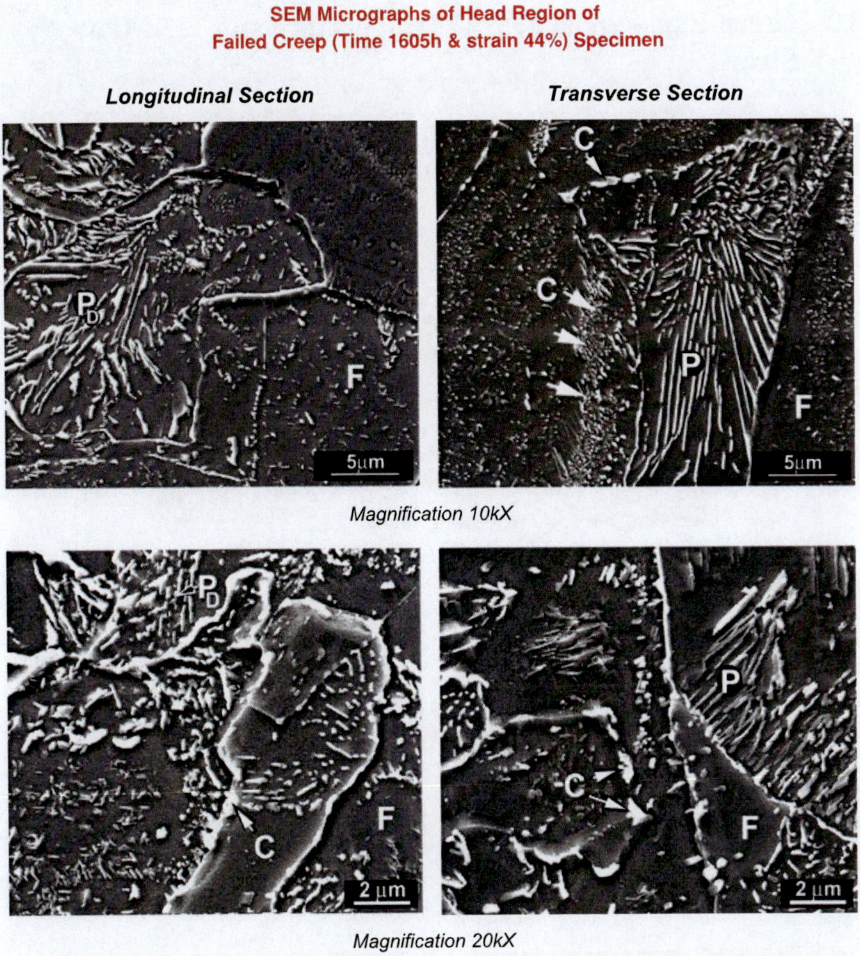

Fig. 4.40 SEM-SE micrographs of **longitudinal and transverse sections** of P22 (2.25Cr–1Mo) steel after <u>*creep failed*</u> at 550 °C/115 MPa for 1605 h, showing inhomogeneous distribution of carbide (C) particles at the grain boundaries, pearlite colonies (P) and discrete pearlite (P_D) in ferrite matrix

4.3 Creep Exposure at 575 °C Temperature and 115 MPa Stress

See Figs. 4.41, 4.42, 4.43, 4.44, 4.45, 4.46, 4.47, 4.48, 4.49, 4.50, 4.51, 4.52, 4.53, 4.54, 4.55, 4.56, 4.57, 4.58, 4.59, 4.60, 4.61 and 4.62.

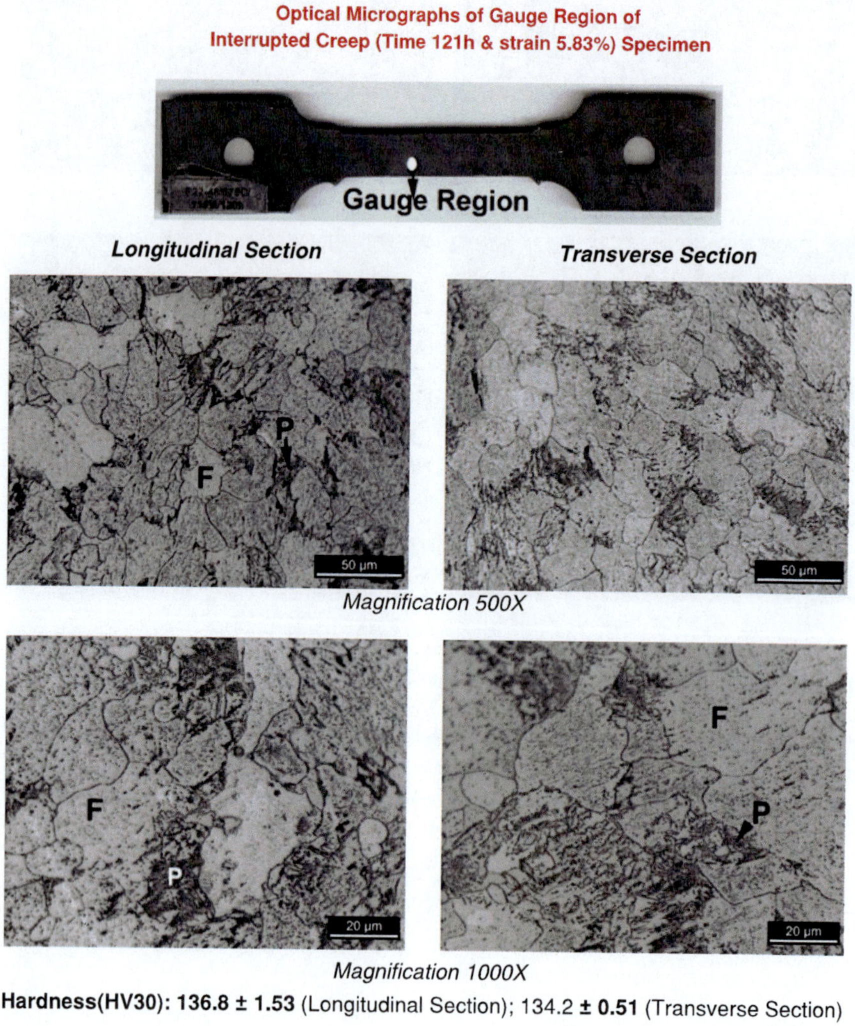

Fig. 4.41 Optical micrographs of **longitudinal and transverse sections** of P22 (2.25Cr–1Mo) steel after creep exposed at 575 °C/115 MPa for 121 h, showing inhomogeneous distribution of ferrite (F) and pearlite (P) structures

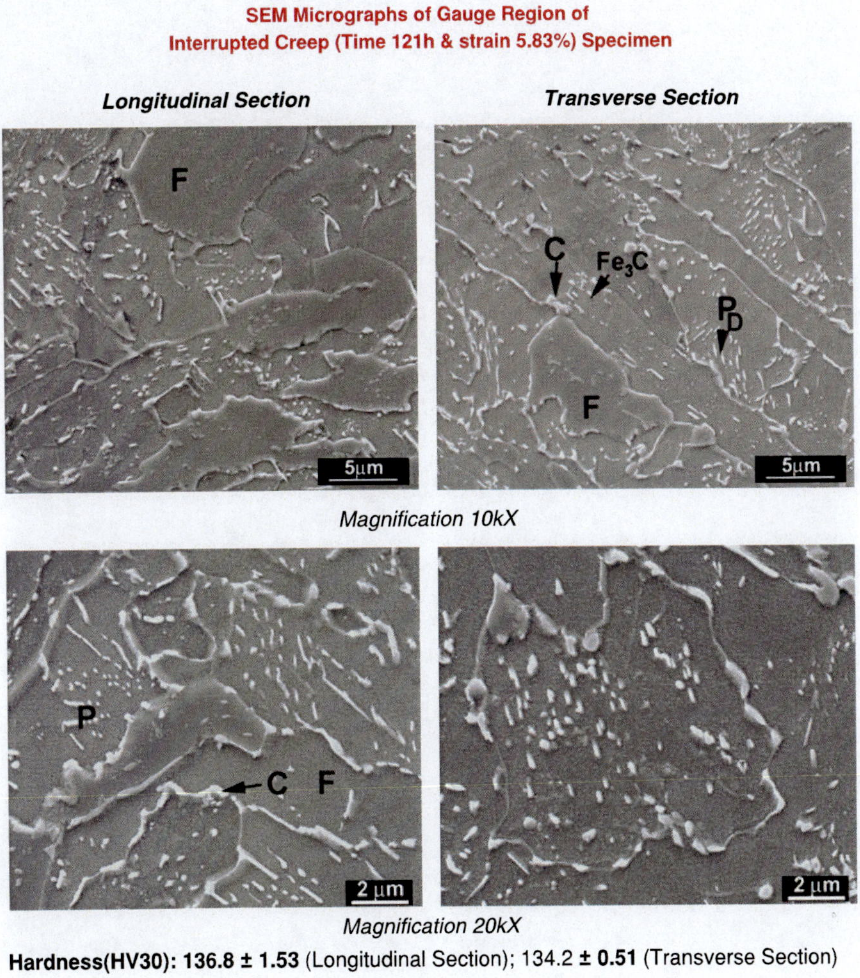

Fig. 4.42 SEM-SE micrographs of **longitudinal and transverse sections** of P22 (2.25Cr–1Mo) steel after creep exposed at 575 °C/115 MPa for 121 h, explaining distribution of pearlite (P) colonies, discrete pearlite (P_D), (Fe, Cr, Mo)-rich carbide (C) at grain boundary and F_3C in grain interior of ferrite (F) matrix

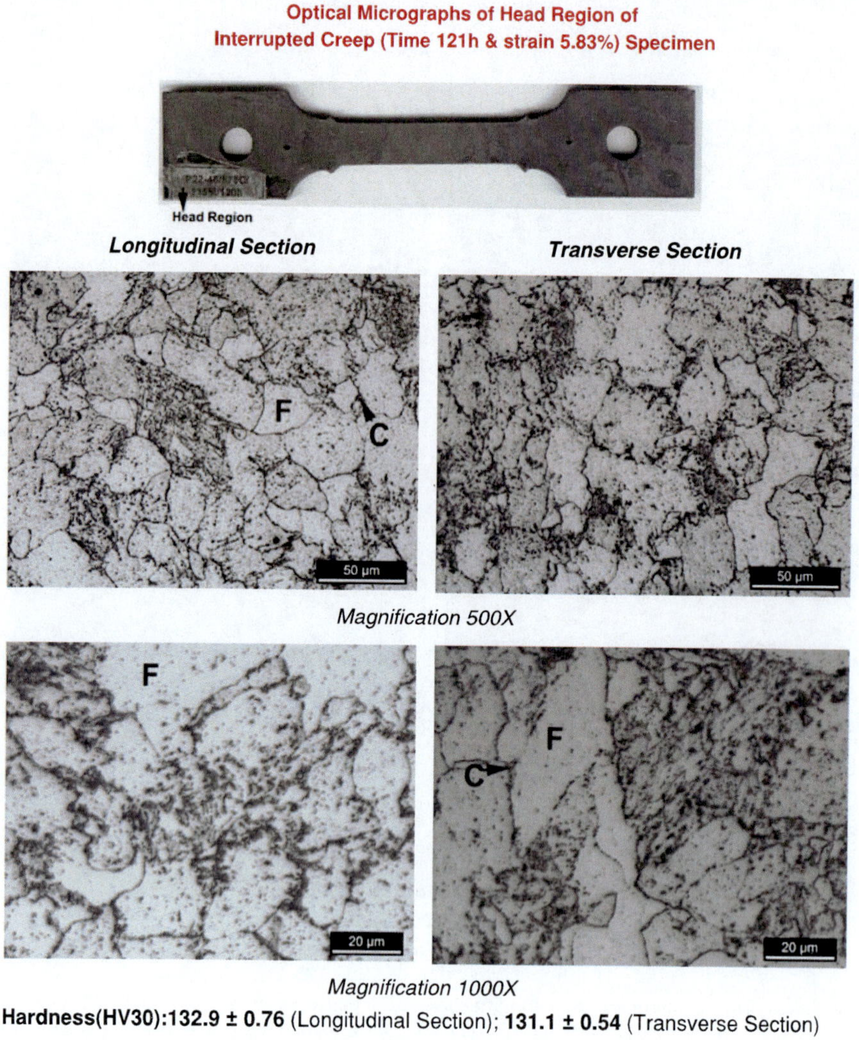

Fig. 4.43 Optical micrographs of **longitudinal and transverse sections** of P22 (2.25Cr–1Mo) steel after creep exposed at 575 °C/115 MPa for 121 h, showing carbide (C) and pearlite in ferrite (F) matrix

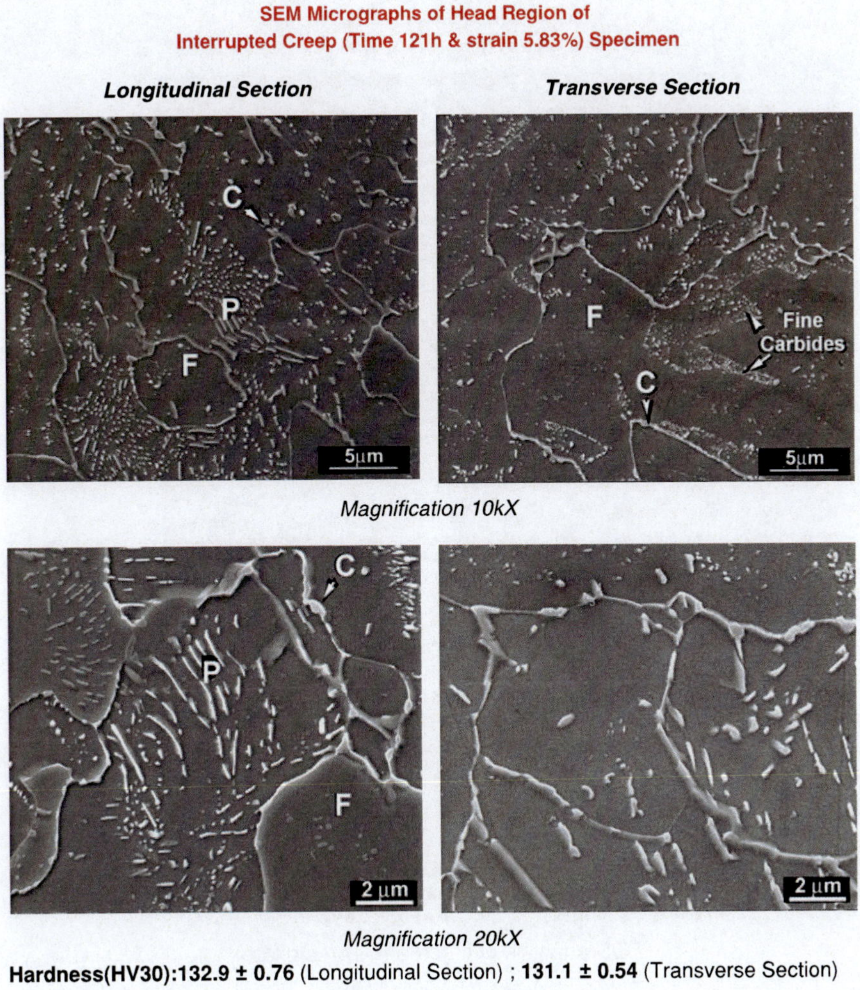

Fig. 4.44 SEM-SE micrographs of **longitudinal and transverse sections** of P22 (2.25Cr–1Mo) steel after creep exposed at 575 °C/115 MPa for 121 h, explaining inhomogeneous distribution of coarse carbides (C) at grain boundaries, fine carbides in grain interior, and pearlite (P) colonies in ferrite (F) matrix

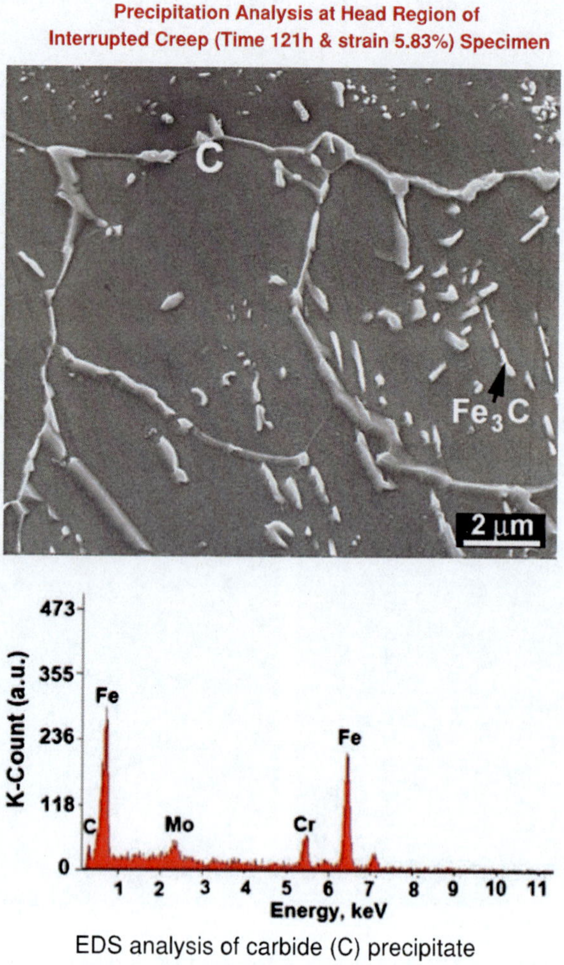

Fig. 4.45 SEM-SE micrographs of (magnification 20 kX) of ***transverse section*** of P22 (2.25Cr–1Mo) steel after creep exposed at 575 °C/115 MPa for 121 h, showing SEM-EDS analysis distribution of (Fe, Cr, Mo)-rich carbides (C) at grain boundaries and Fe_3C in grain interior of ferrite matrix

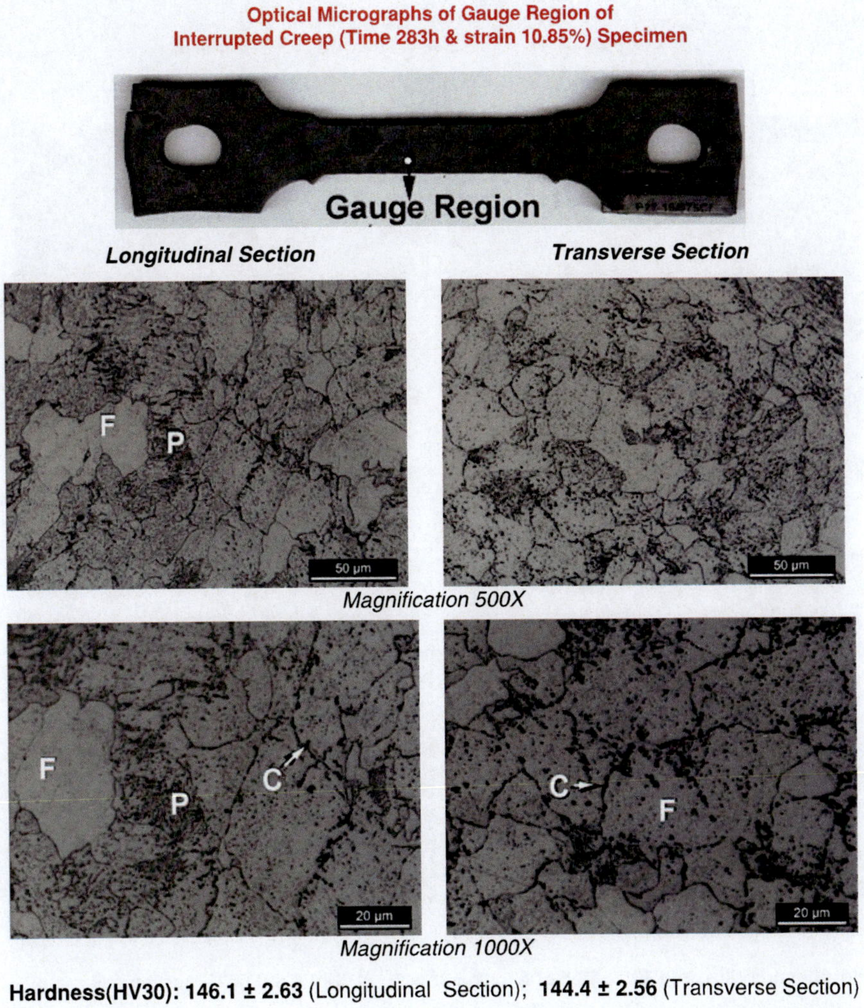

Fig. 4.46 Optical micrographs of **longitudinal and transverse sections** of P22 (2.25Cr–1Mo) steel after creep exposed at 575 °C/115 MPa for 283 h, showing inhomogeneous distribution pearlite (P) and carbides (C) in ferrite (F) matrix

SEM Micrographs of Gauge Region of Interrupted Creep (Time 283h & strain 10.85%) Specimen

Longitudinal Section ***Transverse Section***

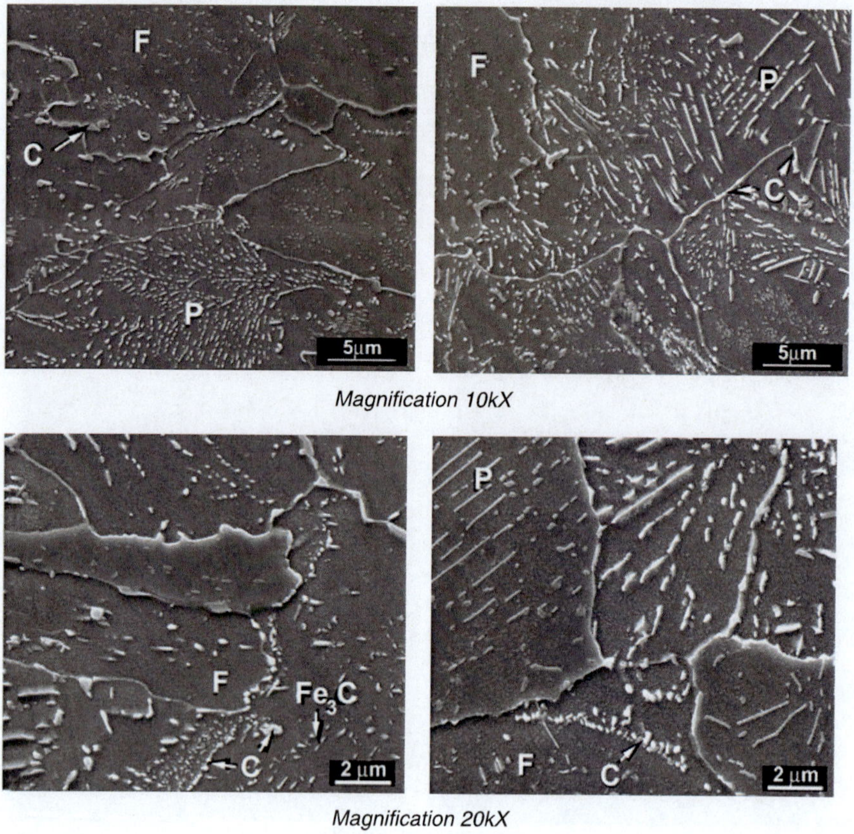

Hardness(HV30): 146.1 ± 2.63 (Longitudinal Section); **144.4 ± 2.56** (Transverse Section)

Fig. 4.47 SEM-SE micrographs of **longitudinal and transverse sections** of P22 (2.25Cr–1Mo) steel after creep exposed at 575 °C/115 MPa for 283 h, explaining inhomogeneous distribution of pearlite (P) and carbides (C) at grain boundaries and grain interior in ferrite (F) matrix

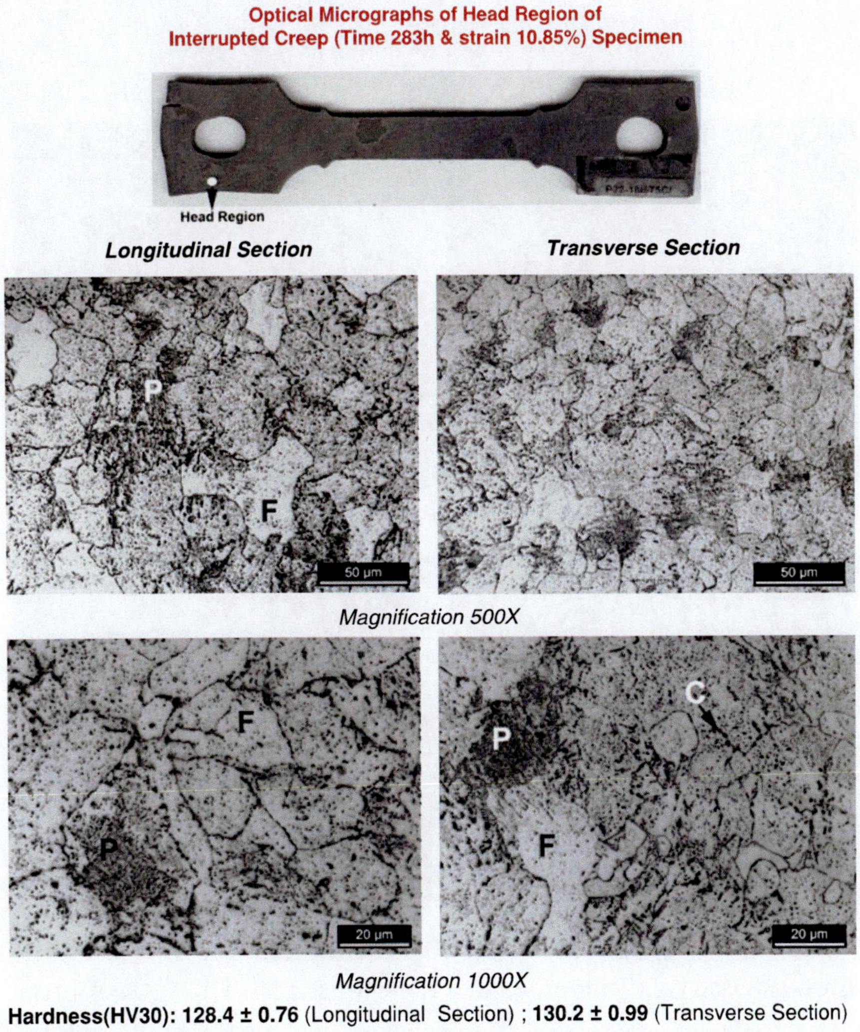

Fig. 4.48 Optical micrographs of **longitudinal and transverse sections** of P22 (2.25Cr–1Mo) steel after creep exposed at 575 °C/115 MPa for 283 h, showing pearlite (P) colonies and carbides (C) in ferrite (F) matrix

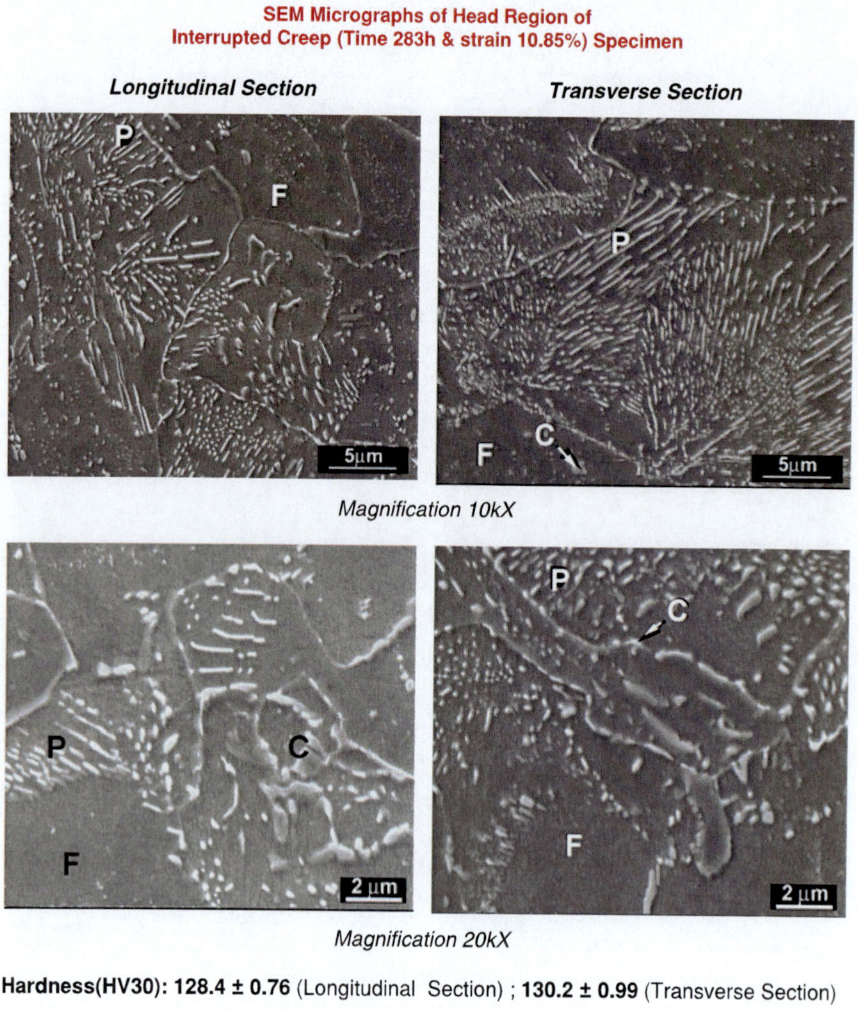

Fig. 4.49 SEM-SE micrographs of **longitudinal and transverse sections** of P22 (2.25Cr–1Mo) steel after creep exposed at 575 °C/115 MPa for 283 h, showing inhomogeneous distribution of pearlite (P) and carbides (C) in ferrite (F) matrix

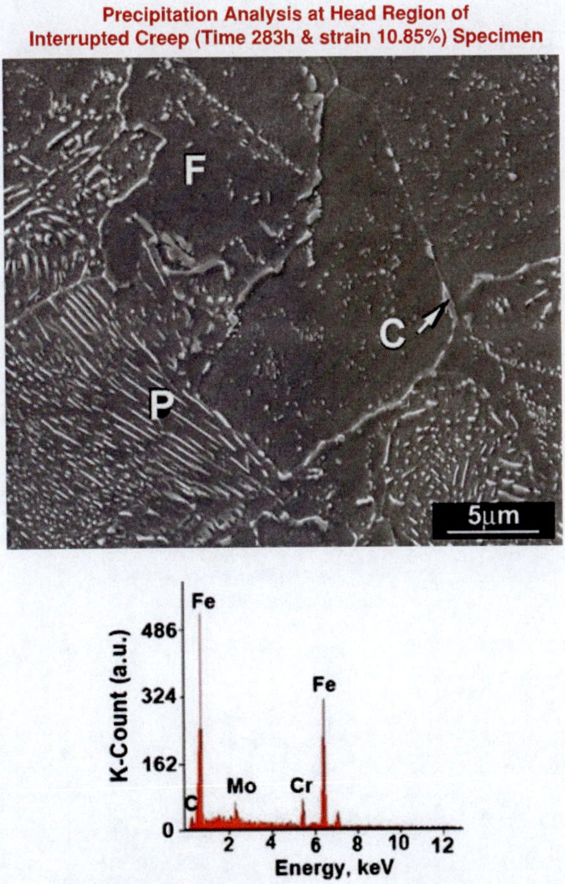

EDS analysis of carbide (C) precipitate

Fig. 4.50 SEM-SE micrograph (magnification 10 kX) of **longitudinal sections** of P22 (2.25Cr–1Mo) steel after creep exposed at 575 °C/115 MPa for 283 h, showing SEM–EDS analysis of (Fe, Cr, Mo)-rich carbide (C) precipitates at grain boundaries and pearlite (P) colonies in ferrite (F) matrix

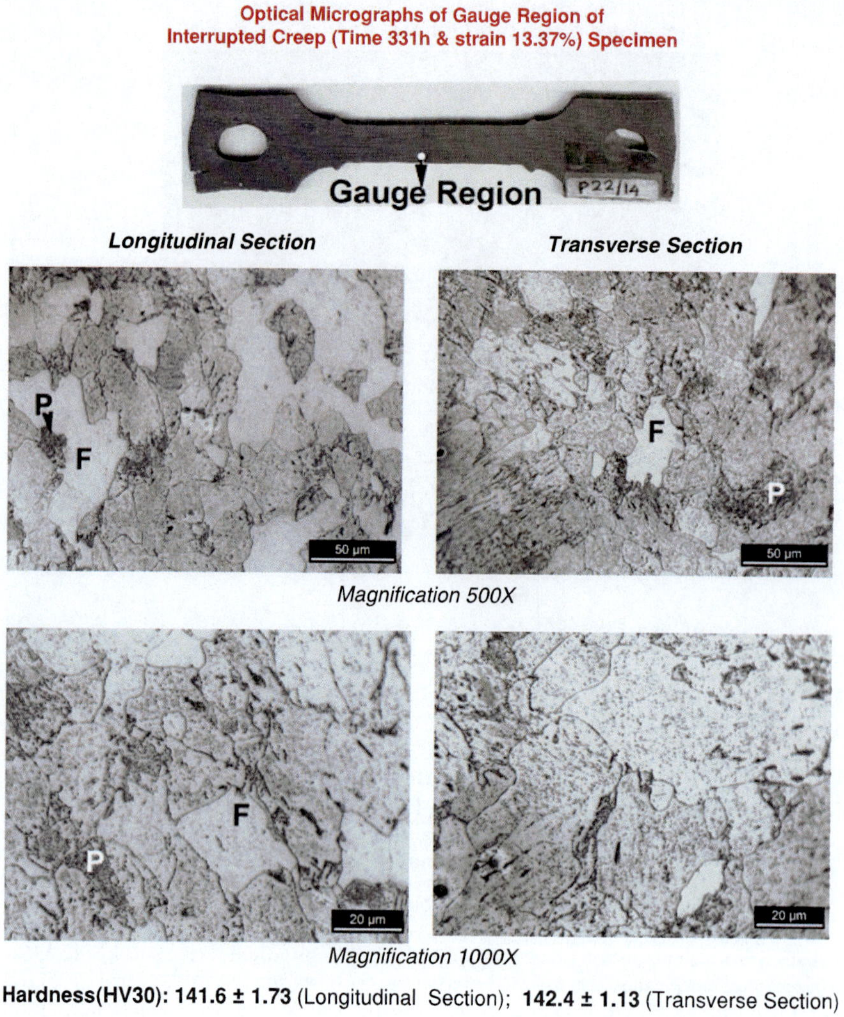

Fig. 4.51 Optical micrographs of **longitudinal and transverse sections** of P22 (2.25Cr–1Mo) steel after creep exposed at 575 °C/115 MPa for 331 h, showing inhomogeneous distribution of ferrite (F) and pearlite (P) structures

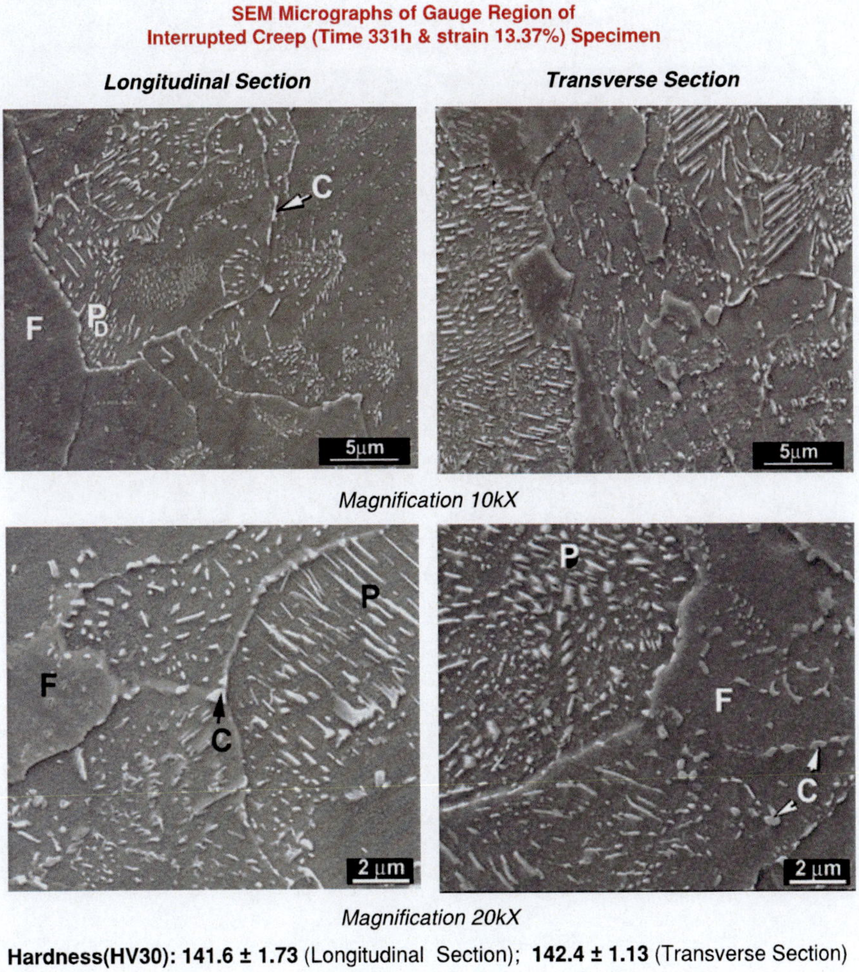

Fig. 4.52 SEM-SE micrographs of **longitudinal and transverse sections** of P22 (2.25Cr–1Mo) steel after creep exposed at 575 °C/115 MPa for 331 h, showing inhomogeneous distribution of pearlite (P) colonies, discrete pearlite (P_D) and carbides (C) in ferrite (F) matrix

Precipitation Analysis at Gauge Region of Interrupted Creep (Time 331h & strain 13.37%) Specimen

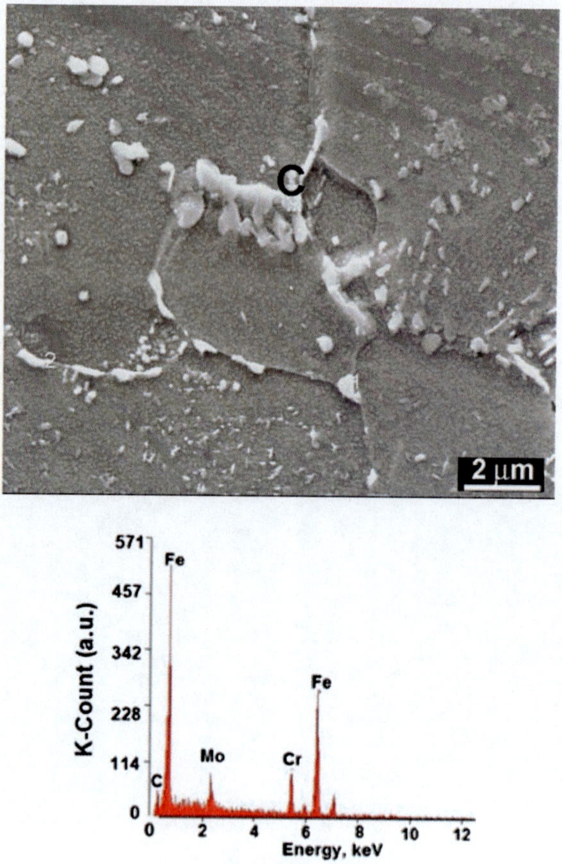

EDS analysis of C precipitates

Fig. 4.53 SEM-SE micrograph (magnification 20 kX) of **transverse sections** of P22 (2.25Cr–1Mo) steel after creep exposed at 575 °C/115 MPa for 331 h, showing carbide (C) precipitates at grain interiors and grain boundaries in ferrite matrix

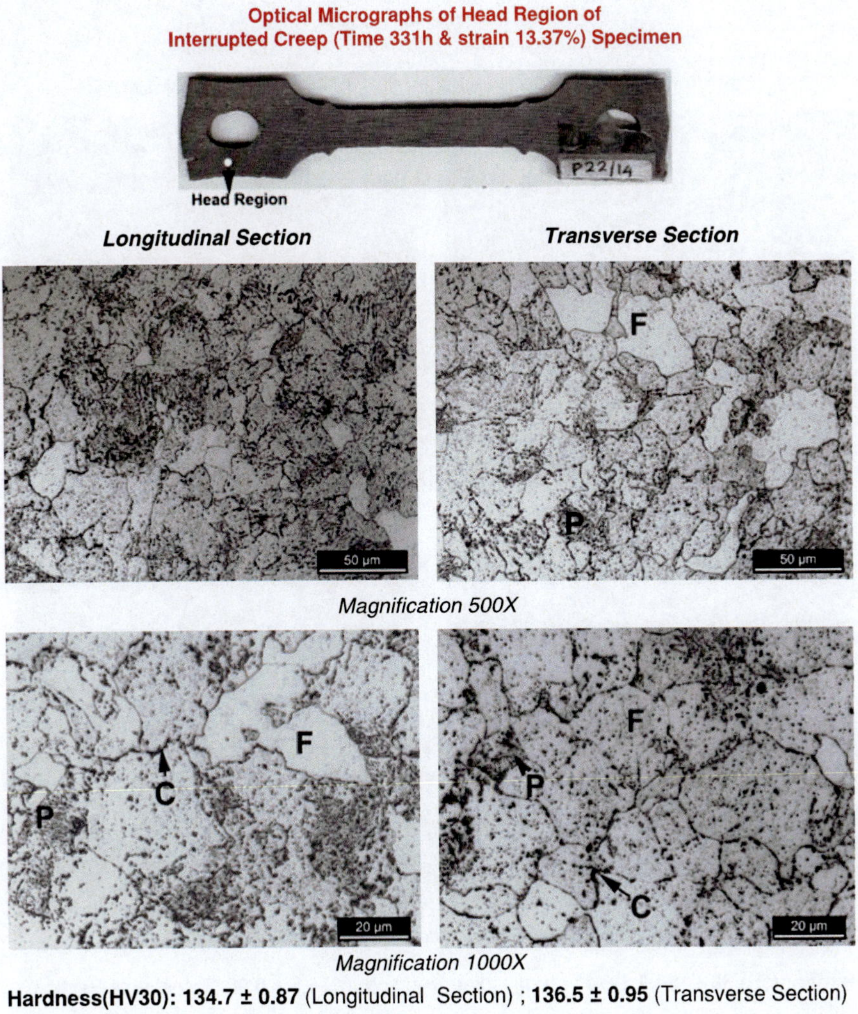

Fig. 4.54 Optical micrographs of **longitudinal and transverse sections** of P22 (2.25Cr–1Mo) steel after creep exposed at 575 °C/115 MPa for 331 h, showing inhomogeneous distribution of carbides (C) at grain boundary and pearlite (P) colonies in ferrite (F) matrix

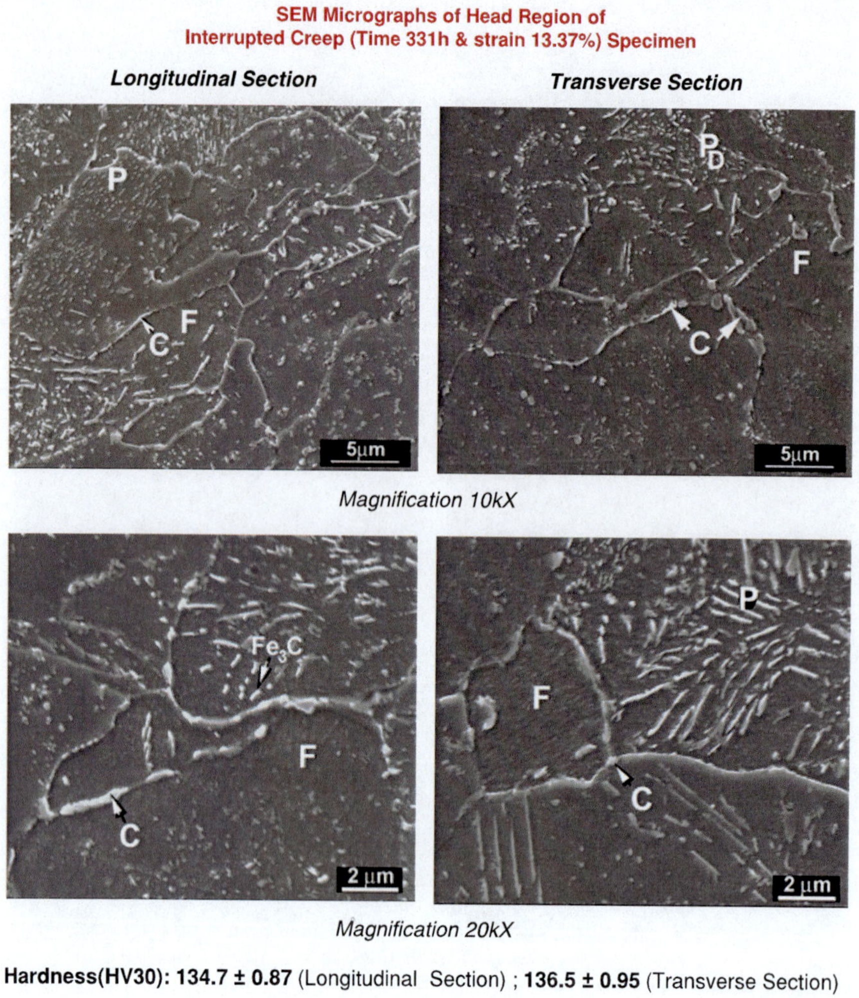

Fig. 4.55 SEM-SE micrographs of **longitudinal and transverse sections** of P22 (2.25Cr–1Mo) steel after creep exposed at 575 °C/115 MPa for 331 h, explaining pearlite (P) colonies, discrete pearlite (P_D), Fe_3C and carbides (C) in ferrite (F) matrix

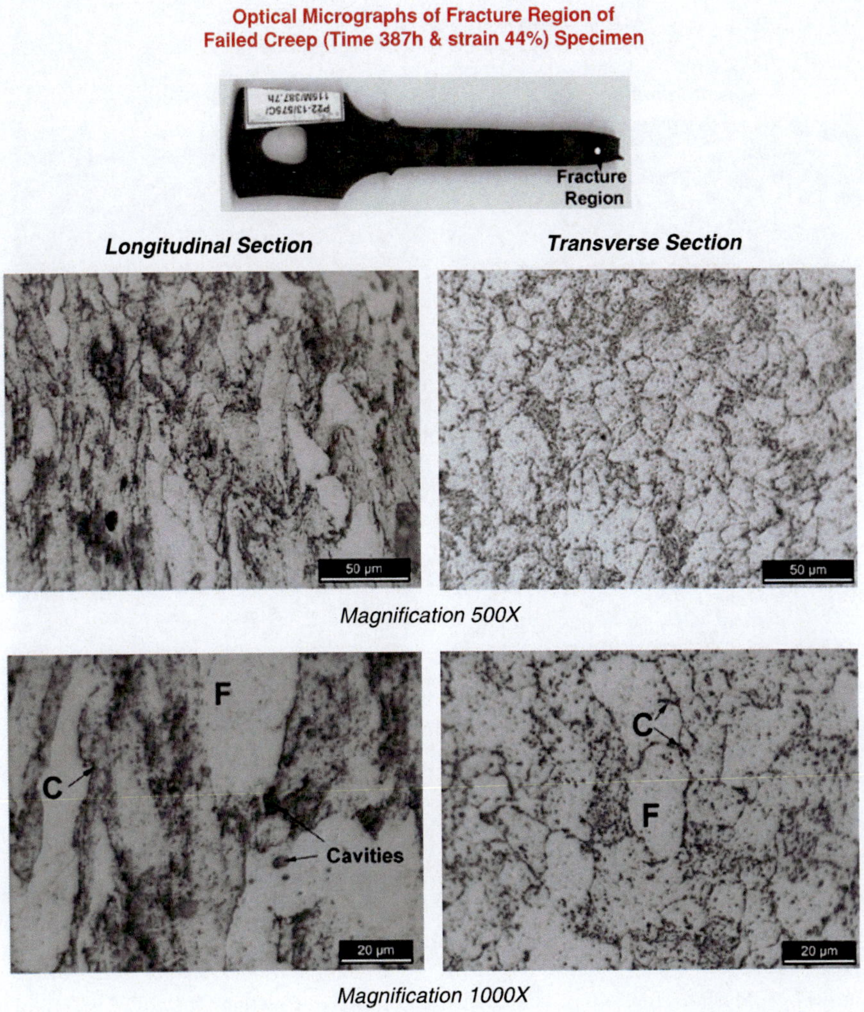

Fig. 4.56 Optical micrographs of **longitudinal and transverse sections** of P22 (2.25Cr–1Mo) steel after *creep failed* at 575 °C/115 MPa for 387 h, showing cavity formation in ferrite (F) matrix and carbide (C) precipitates at grain boundaries

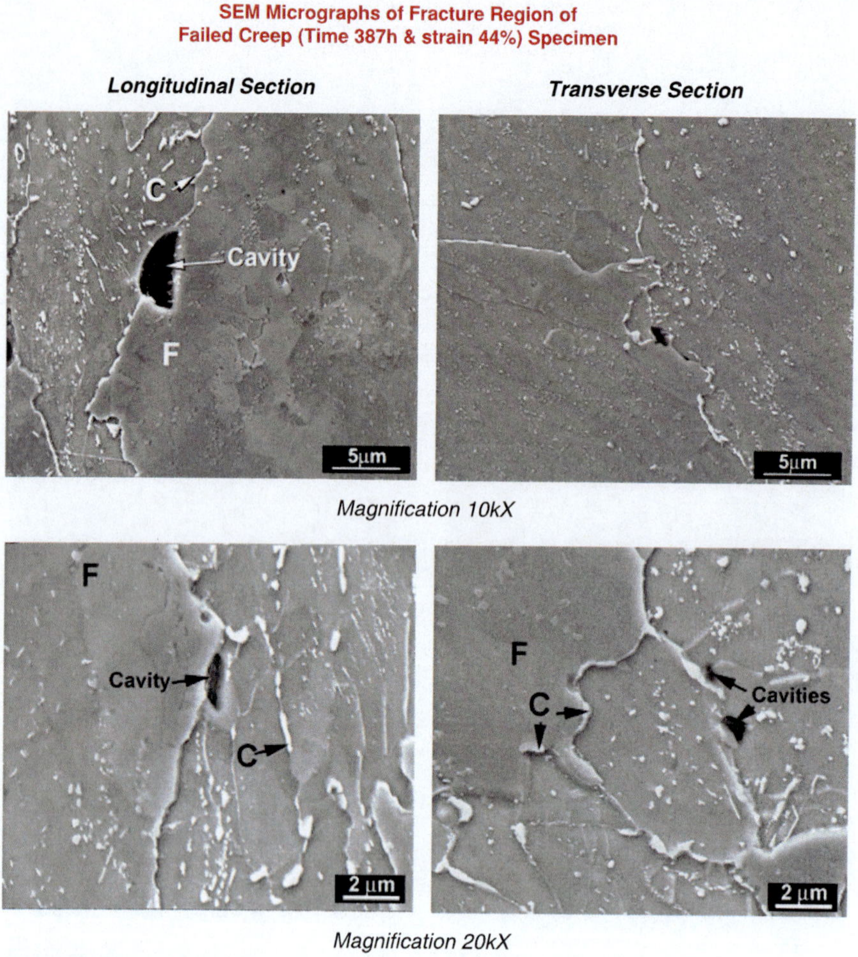

Fig. 4.57 SEM-SE micrographs of **longitudinal and transverse sections** of P22 (2.25Cr–1Mo) steel after creep failed at 575 °C/115 MPa for 387 h, showing cavity and carbide (C) precipitates at grain boundaries in ferrite (F) matrix

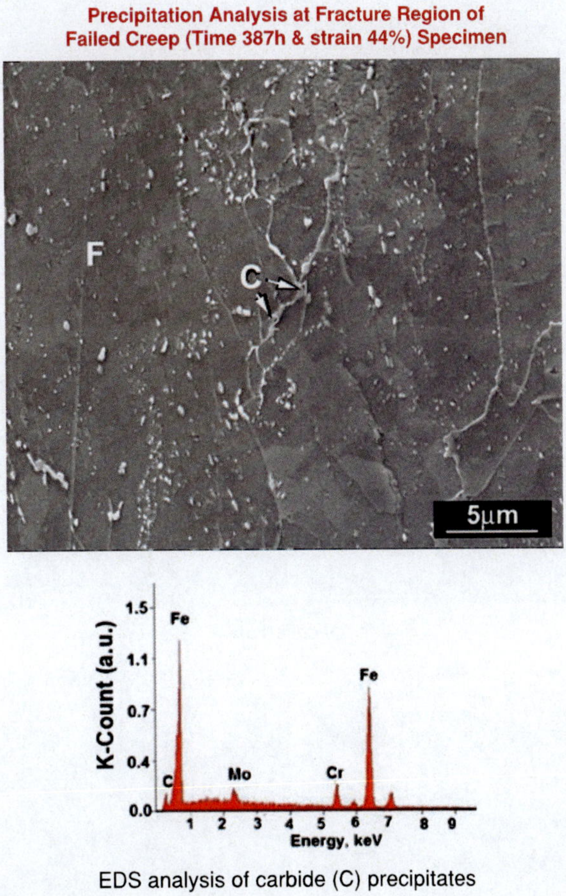

Fig. 4.58 SEM-SE micrographs of **longitudinal section** of P22 (2.25Cr–1Mo) steel after *creep failed* at 575 °C/115 MPa for 387 h, showing SEM-EDS analysis of (Fe, Cr, Mo)-rich carbide (C) precipitates at grain boundaries in ferrite (F) matrix

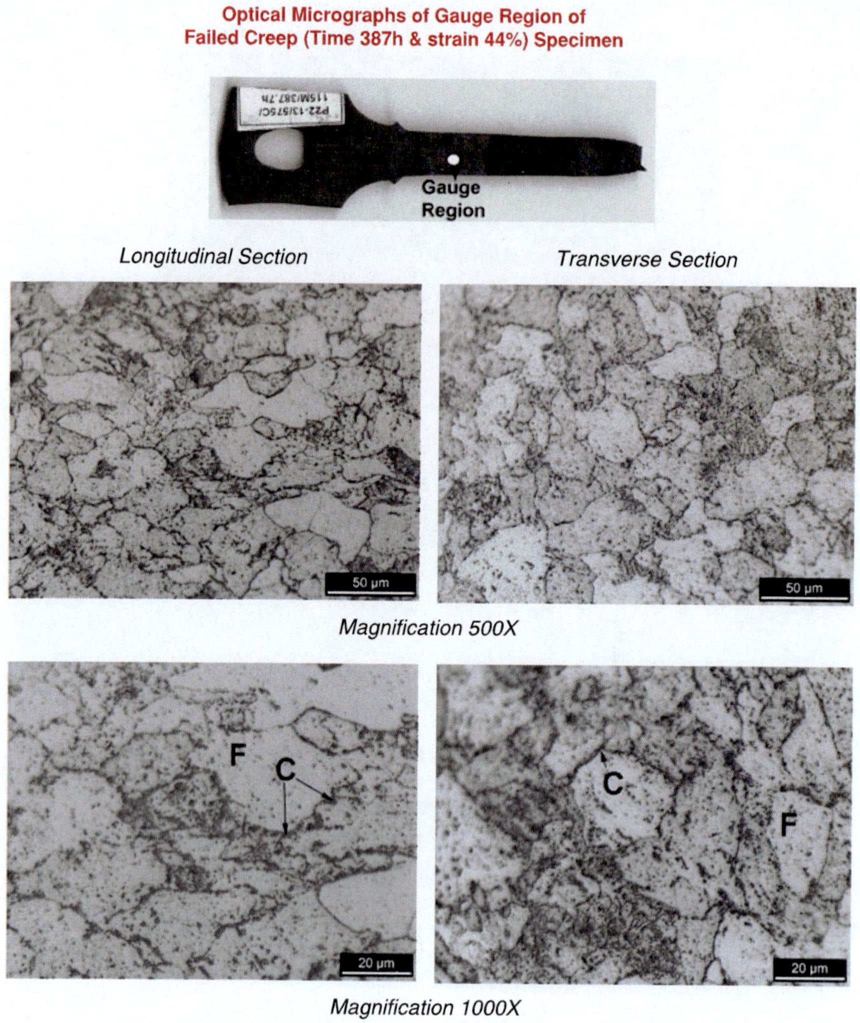

Fig. 4.59 Optical micrographs of longitudinal and transverse sections of P22 (2.25Cr–1Mo) steel after creep failed at 575 °C/115 MPa for 387 h, explaining carbide (C) precipitates at grain boundaries in ferrite (F) matrix

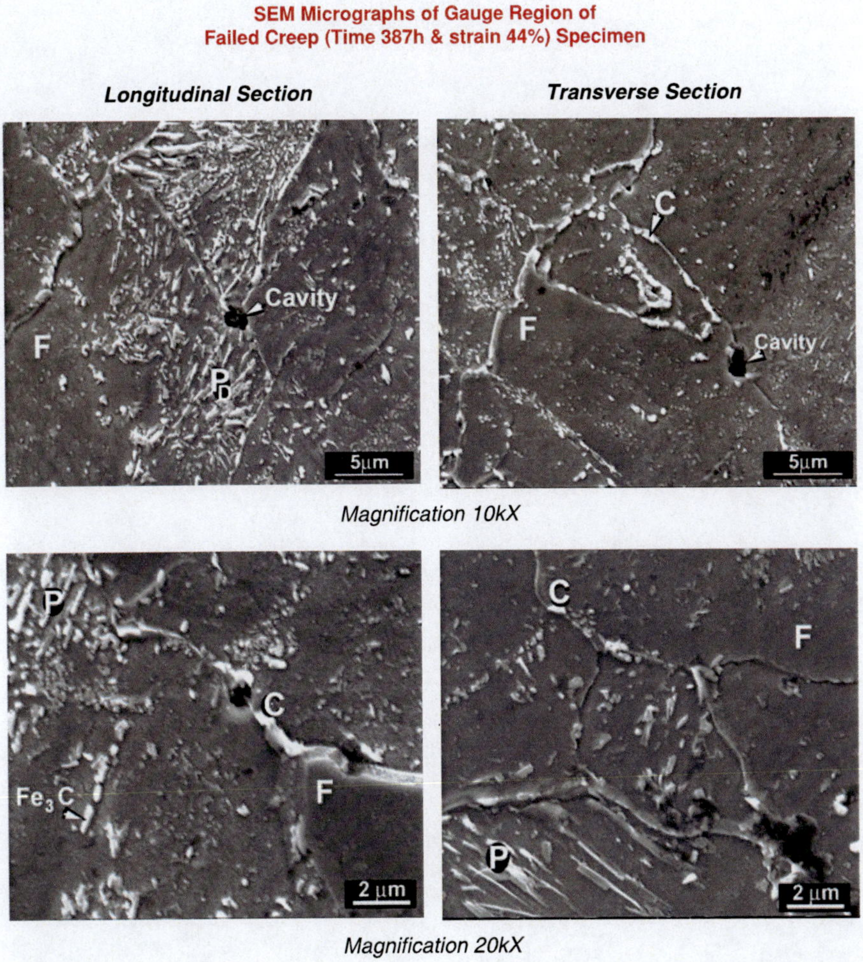

Fig. 4.60 SEM-SE micrographs of **longitudinal and transverse sections** of P22 (2.25Cr–1Mo) steel after creep failed at 575 °C/115 MPa for 387 h, showing cavity and (Fe, Cr, Mo)-rich carbide (C) precipitates at grain boundaries and pearlite (P) colonies and discrete pearlite (P_D) in ferrite (F) matrix

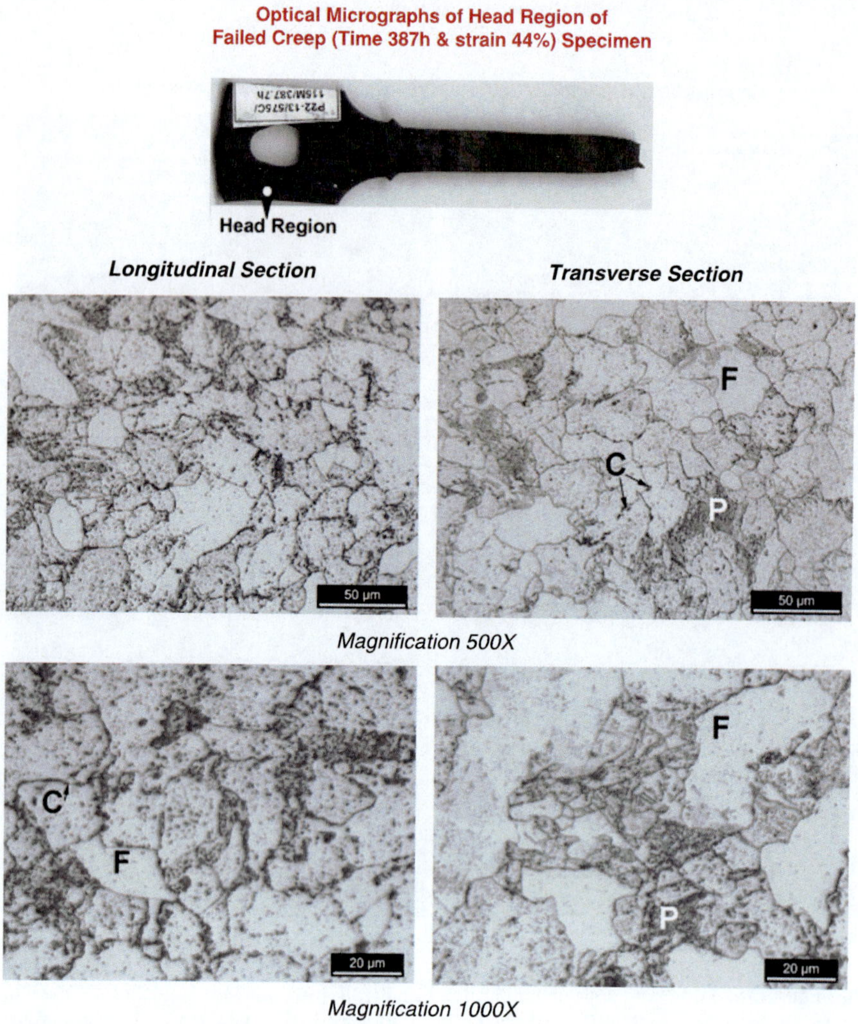

Fig. 4.61 Optical micrographs of **longitudinal and transverse sections** of P22 (2.25Cr–1Mo) steel after *creep failed* at 575 °C/115 MPa for 387 h, showing carbide (C) precipitates at grain boundaries and pearlite (P) in ferrite (F) matrix

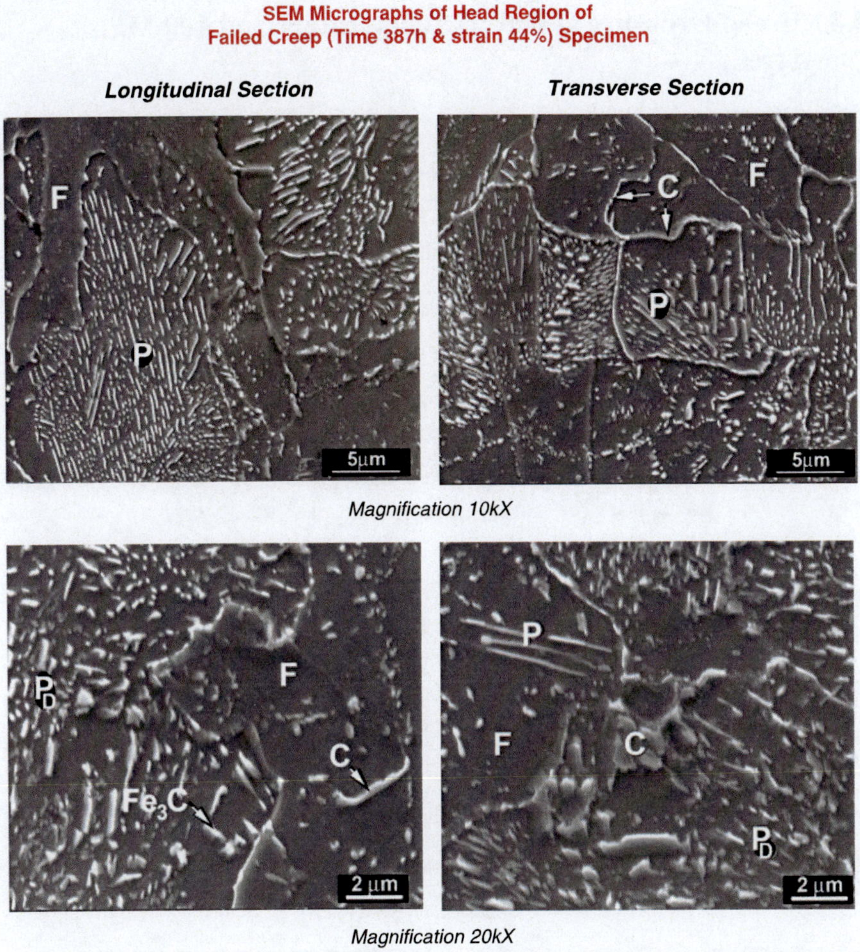

Fig. 4.62 SEM-SE micrographs of **longitudinal and transverse sections** of P22 (2.25Cr–1Mo) steel after *creep failed* at 575 °C/115 MPa for 387 h, showing carbide (C) precipitates at grain boundaries and cementite (Fe_3C), pearlite (P) colonies and discrete pearlite (P_D) in ferrite (F) matrix

4.4 Creep Exposure at 600 °C Temperature and 100 MPa Stress

See Figs. 4.63, 4.64, 4.65, 4.66, 4.67, 4.68, 4.69, 4.70, 4.71, 4.72, 4.73, 4.74, 4.75, 4.76, 4.77, 4.78, 4.79, 4.80, 4.81, 4.82 and 4.83.

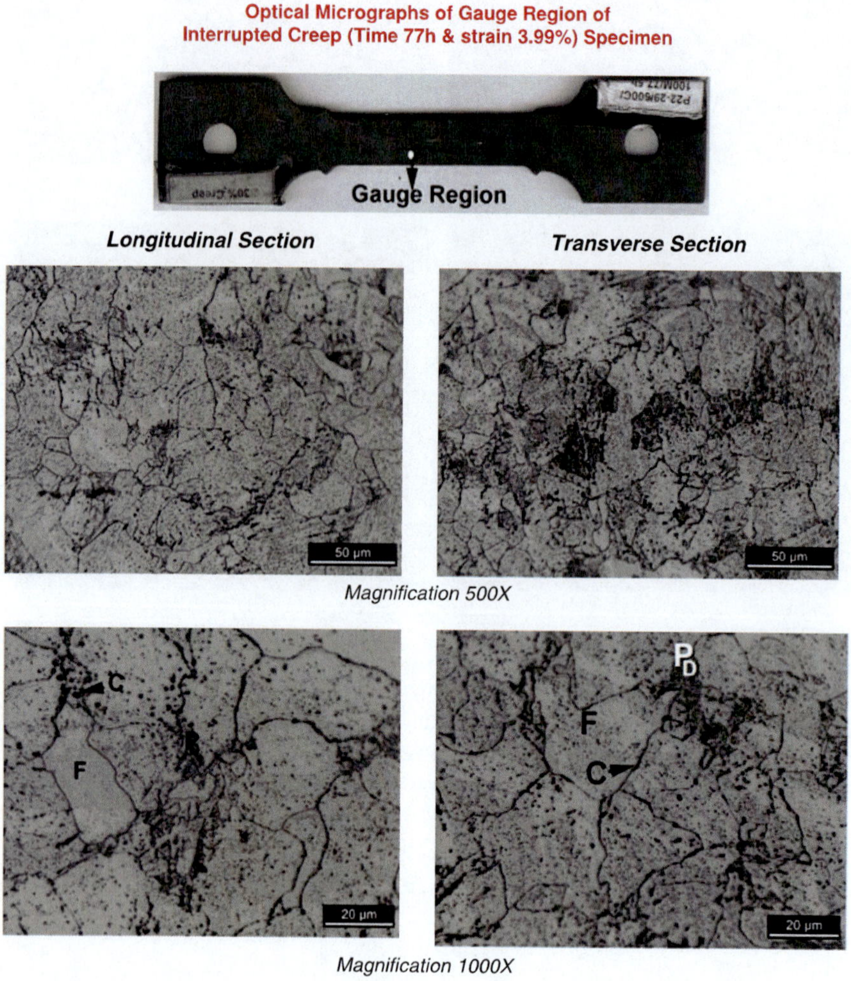

Fig. 4.63 Optical micrographs of **longitudinal and transverse sections** of P22 (2.25Cr–1Mo) steel after creep exposed at 600 °C/100 MPa for 77 h, showing inhomogeneous distribution of discrete pearlite (P_D) and carbides (C) in ferrite (F) matrix

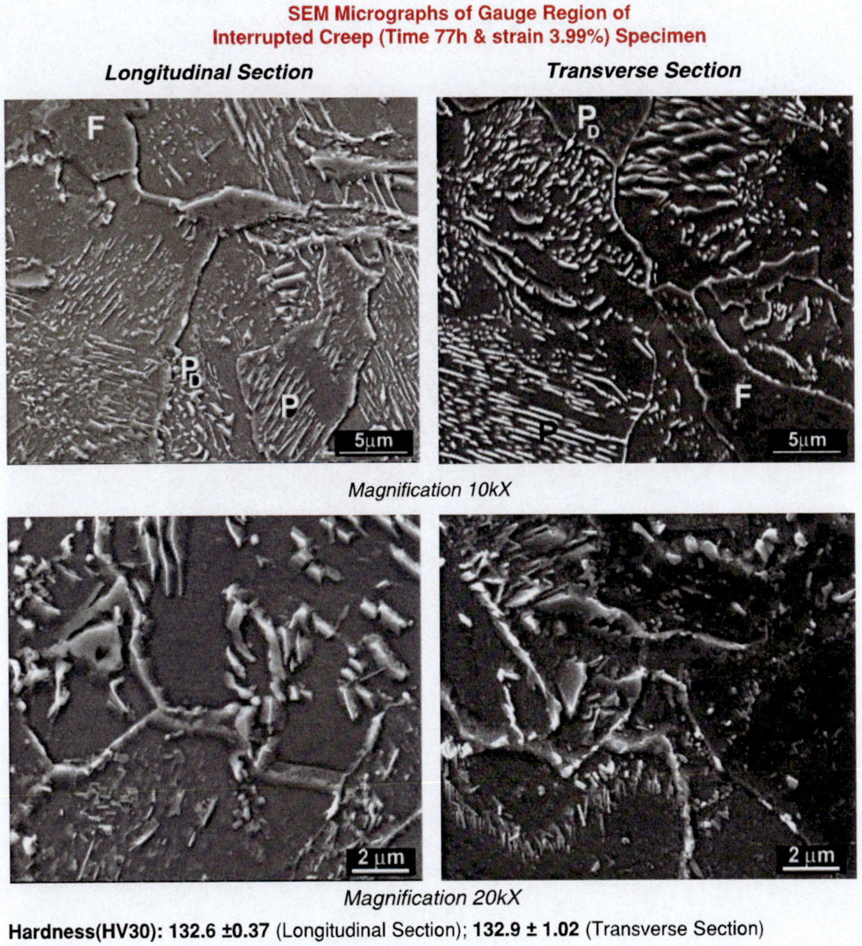

Fig. 4.64 SEM-SE micrographs of **longitudinal and transverse sections** of P22 (2.25Cr–1Mo) steel after creep exposed at 600 °C/100 MPa for 77 h, showing rich carbides, discrete pearlite (P_D) and pearlite (P) colonies in ferrite (F) matrix

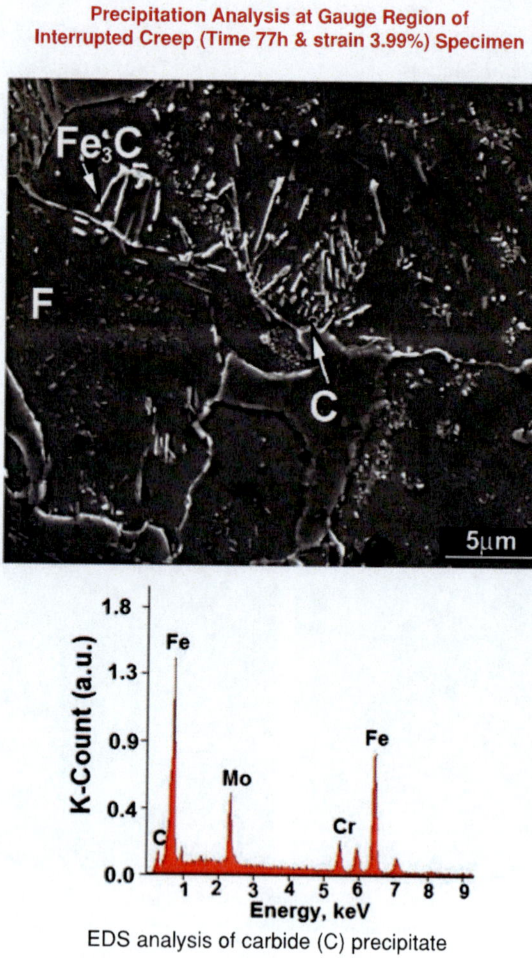

Fig. 4.65 SEM-SE micrograph (magnification 10 kX) of **transverse section** of P22 (2.25Cr–1Mo) steel after creep exposed at 600 °C/100 MPa for 77 h, explaining (Fe, Cr, Mo)-rich carbides (C) and cementite (Fe_3C) as the pearlite colonies in ferrite (F) matrix

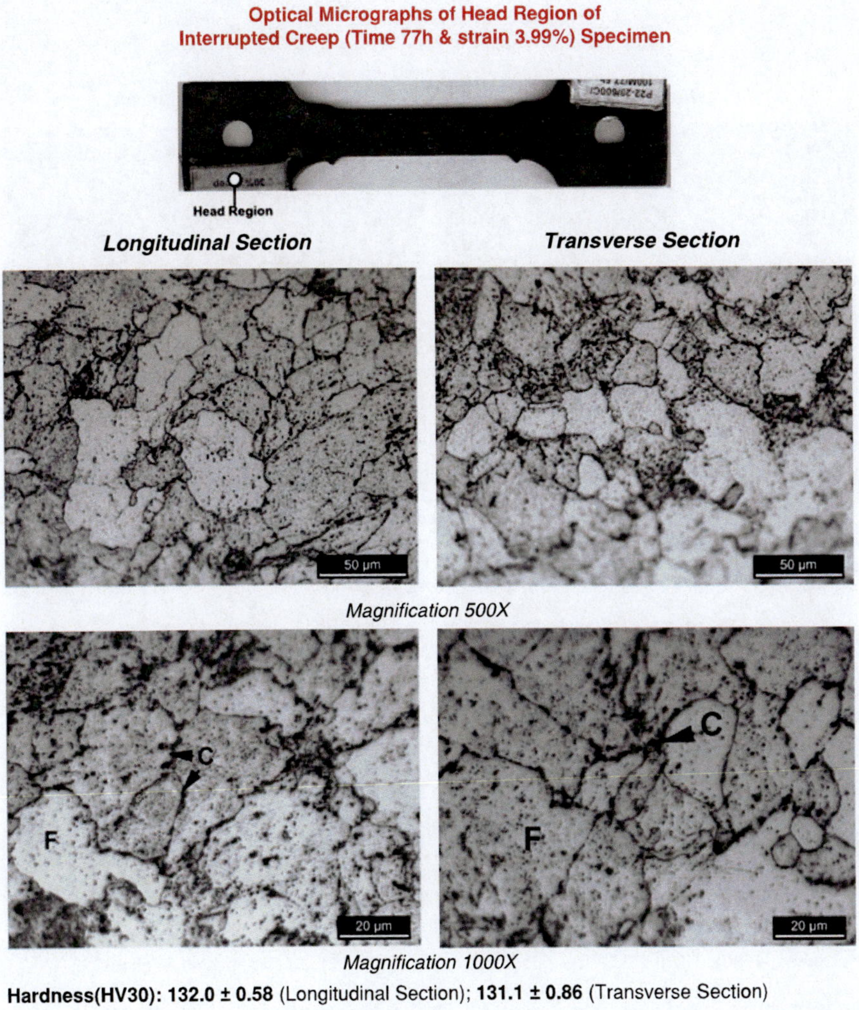

Fig. 4.66 Optical micrographs of **longitudinal and transverse sections** of P22 (2.25Cr–1Mo) steel after creep exposed at 600 °C/100 MPa for 77 h, showing carbides (C) at grain boundaries of ferrite (F) matrix

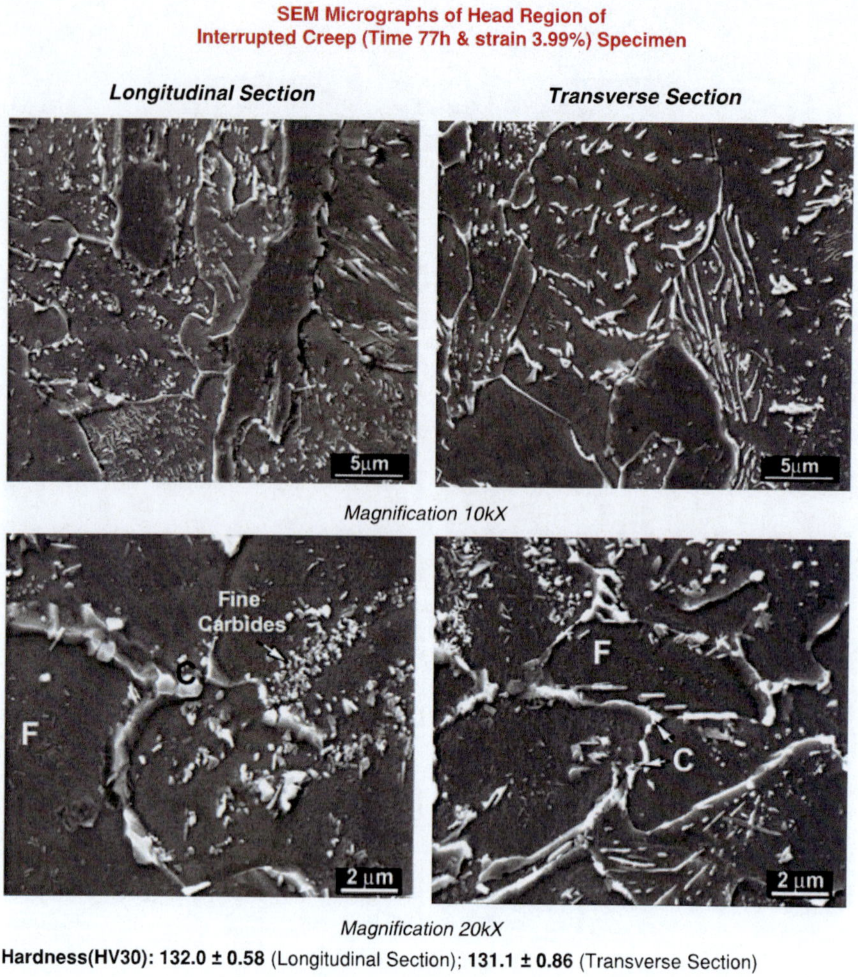

Fig. 4.67 SEM-SE micrographs of **longitudinal and transverse sections** of P22 (2.25Cr–1Mo) steel after creep exposed at 600 °C/100 MPa for 77 h, showing fine carbides in grain interiors and coarse carbide (C) in ferrite (F) matrix

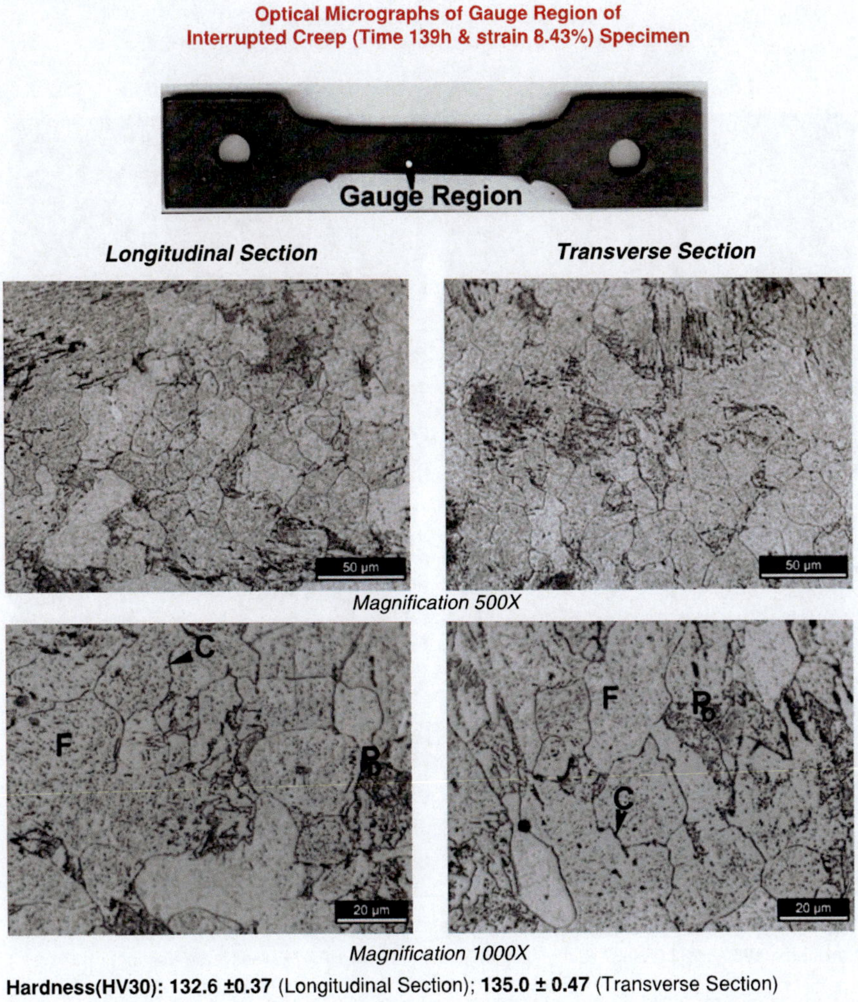

Fig. 4.68 Optical micrographs of **longitudinal and transverse sections** of P22 (2.25Cr–1Mo) steel after creep exposed at 600 °C/100 MPa for 139 h, demonstrating discrete pearlite (P_D) and carbides (C) distributed inhomogeneously in ferrite (F) matrix

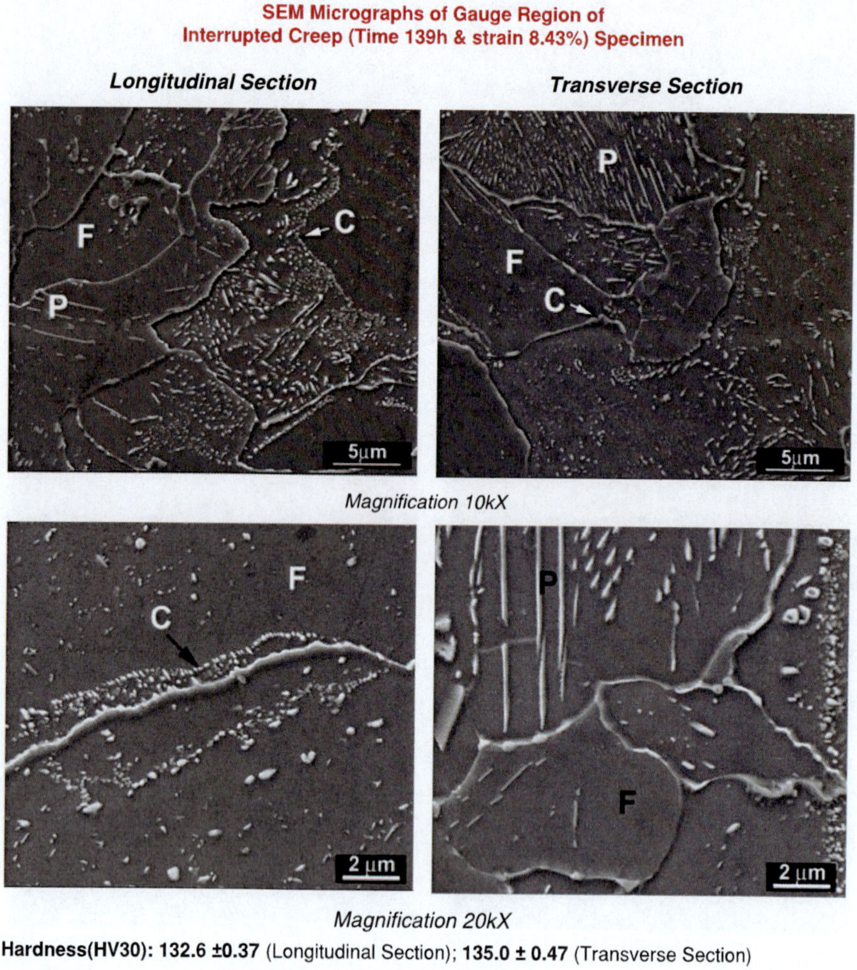

Fig. 4.69 SEM-SE micrographs of **longitudinal and transverse sections** of P22 (2.25Cr–1Mo) steel after creep exposed at 600 °C/100 MPa for 139 h, showing distribution of pearlite (P) colonies, fine carbides at grain interiors and coarse carbide at grain boundaries in ferrite (F) matrix

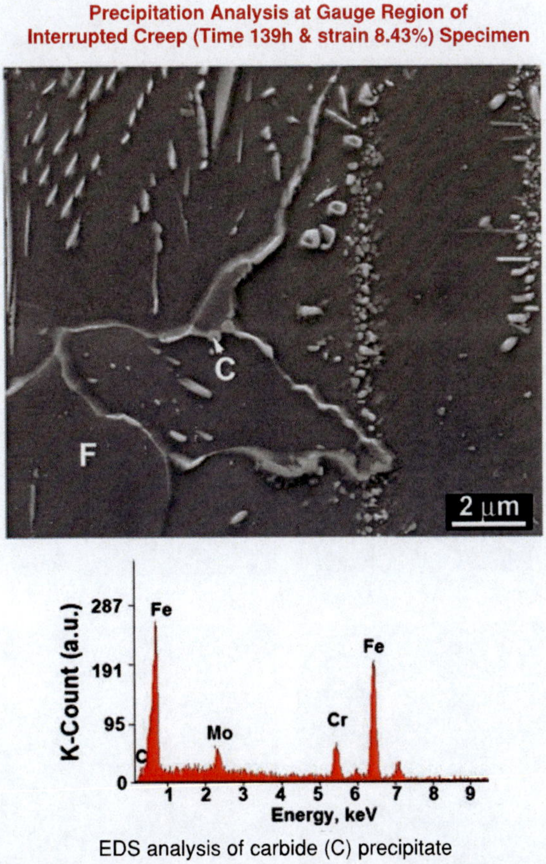

Fig. 4.70 SEM-SE micrograph (magnification 20 kX) of **transverse sections** of P22 (2.25Cr–1Mo) steel after creep exposed at 600 °C/100 MPa for 139 h, explaining EDS analysis of (Fe, Cr, Mo)-rich carbide (C) precipitates at the grain boundaries of ferrite (F) matrix

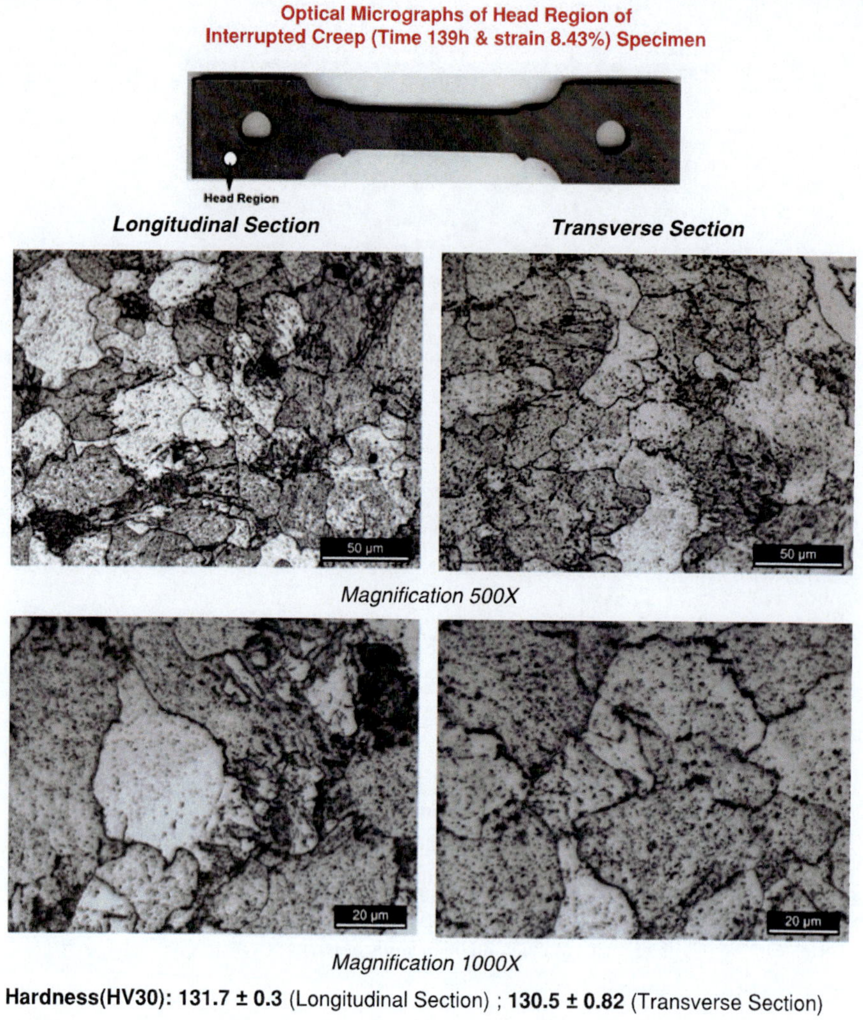

Fig. 4.71 Optical micrographs of **longitudinal and transverse sections** of P22 (2.25Cr–1Mo) steel after creep exposed at 600 °C/100 MPa for 139 h, showing inhomogeneous distribution of pearlite colonies in ferrite matrix

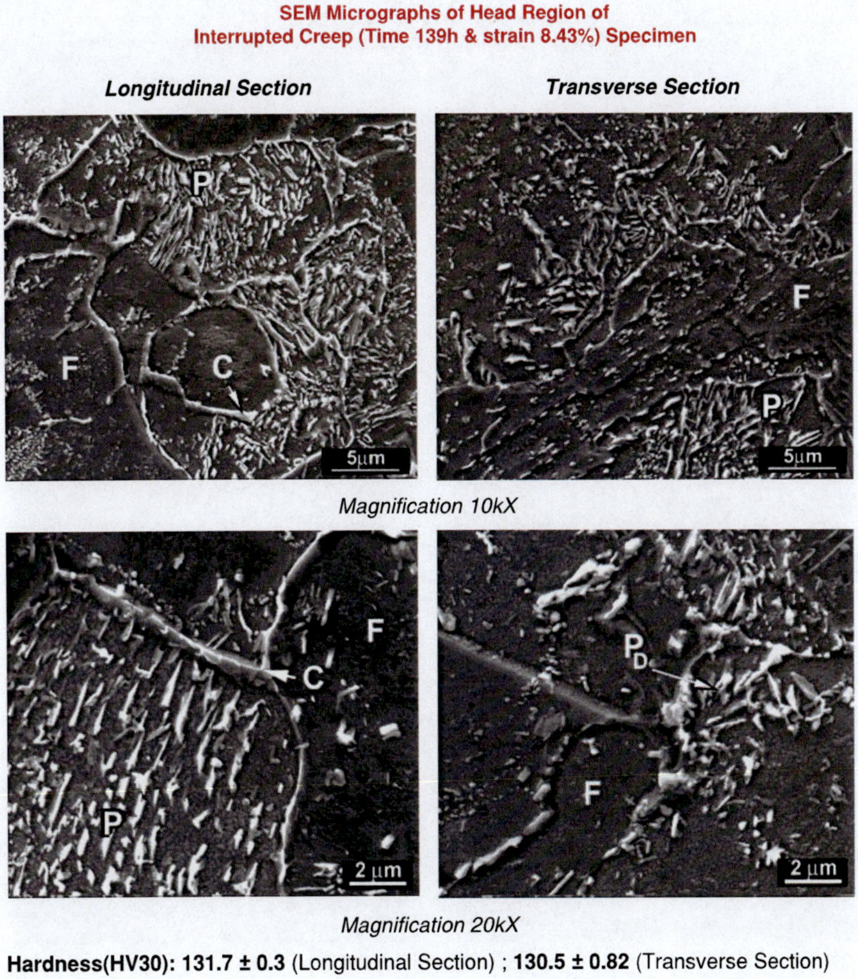

Fig. 4.72 SEM-SE micrographs of **longitudinal and transverse sections** of P22 (2.25Cr–1Mo) steel after creep exposed at 600 °C/100 MPa for 139 h, explaining carbides (C) at grain boundaries and pearlite (P) colonies, discrete pearlite (P_D) in grain interiors of ferrite (F) matrix

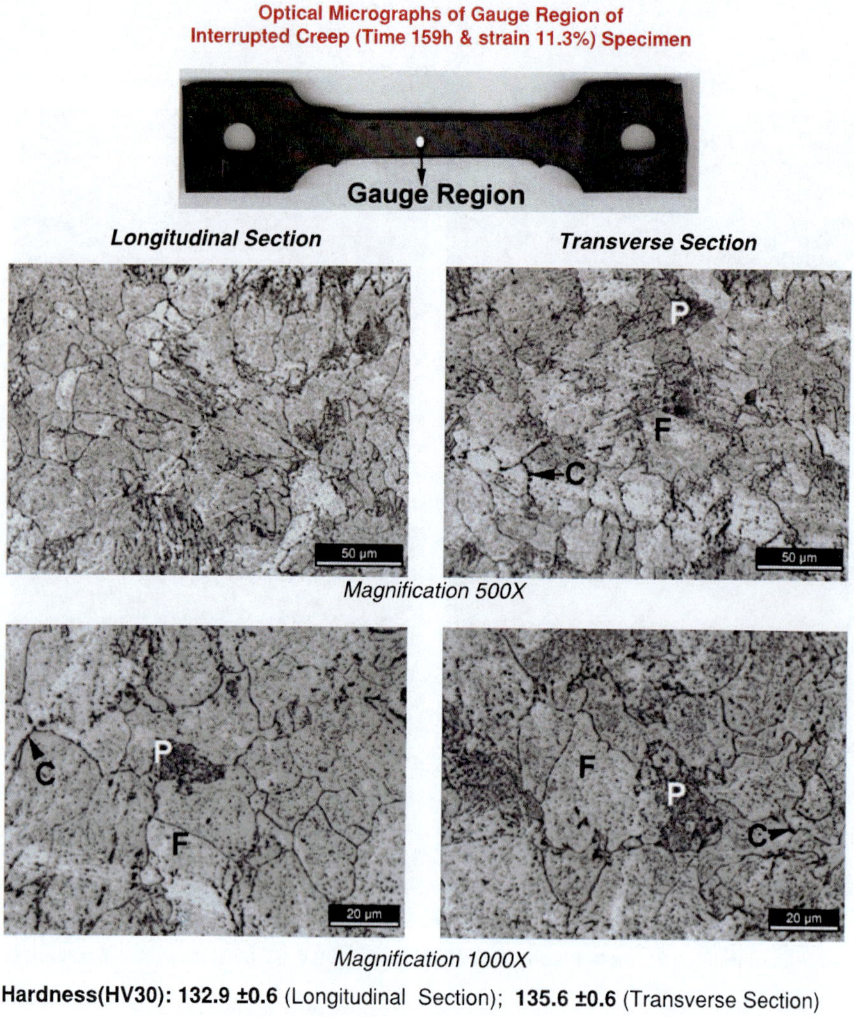

Fig. 4.73 Optical micrographs of **longitudinal and transverse sections** of P22 (2.25Cr–1Mo) steel after creep exposed at 600 °C/100 MPa for 159 h, showing inhomogeneous distribution of carbides (C), ferrite (F) and pearlite (P) structures

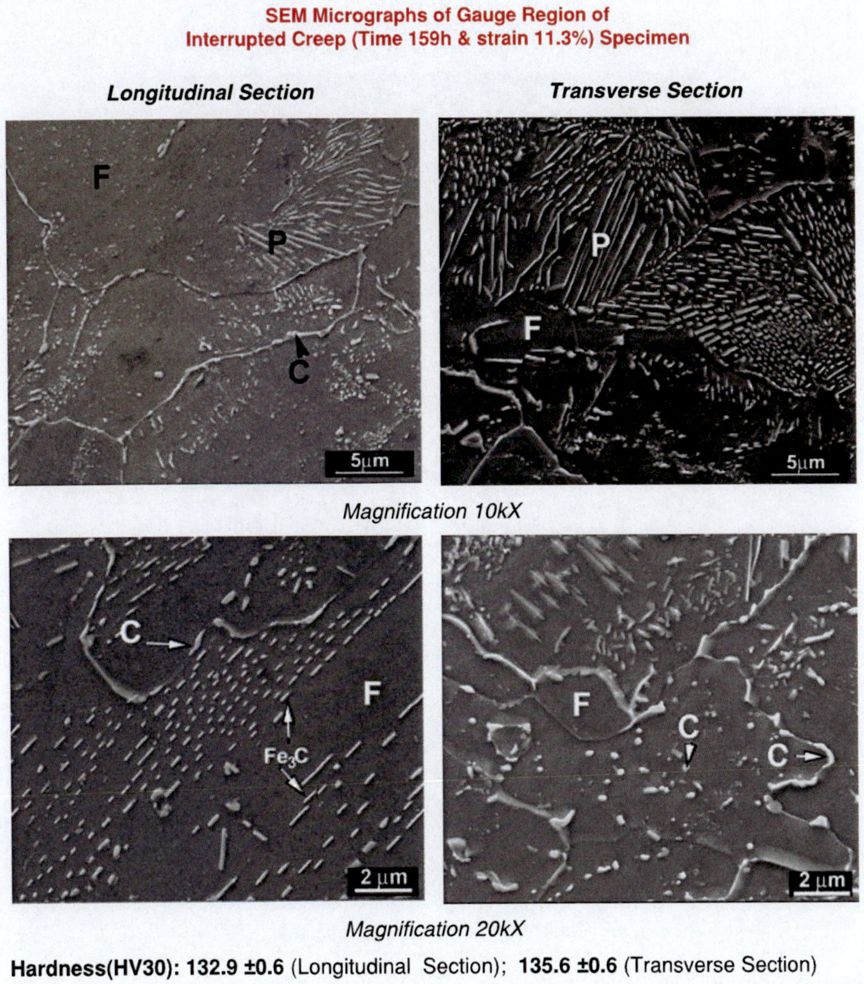

Fig. 4.74 SEM-SE micrographs of **longitudinal and transverse sections** of P22 (2.25Cr–1Mo) steel after creep exposed at 600 °C/100 MPa for 159 h, explaining carbides (C) at grain boundaries and cementite (Fe_3C), pearlite (P) colonies at grain interiors in ferrite (F) matrix

Precipitation Analysis at Gauge Region of
Interrupted Creep (Time 159h & strain 11.3%) Specimen

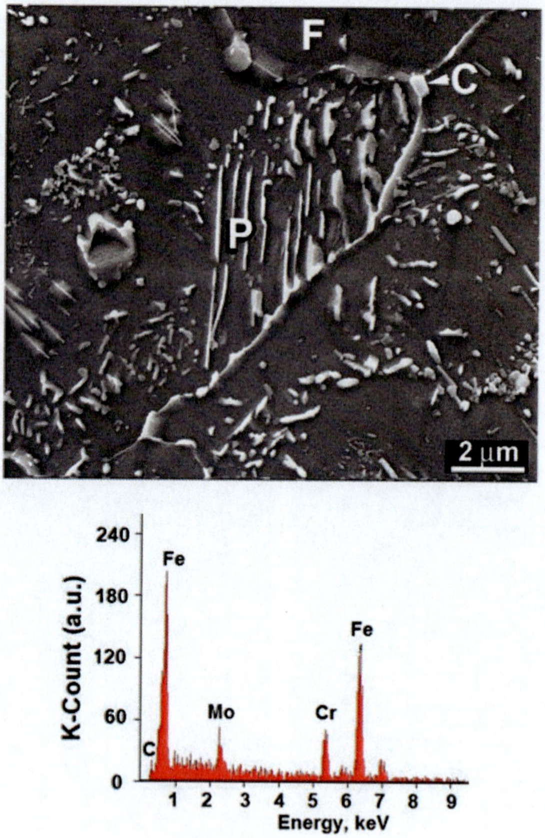

EDS analysis of carbide (C) precipitate

Fig. 4.75 SEM-SE micrograph (magnification 20 kX) of **transverse sections** of P22 (2.25Cr–1Mo) steel after creep exposed at 600 °C/100 MPa for 159 h, explaining EDS analysis of (Fe, Cr, Mo)-rich carbide (C) precipitates at the grain boundaries of ferrite (F) matrix and pearlite colonies

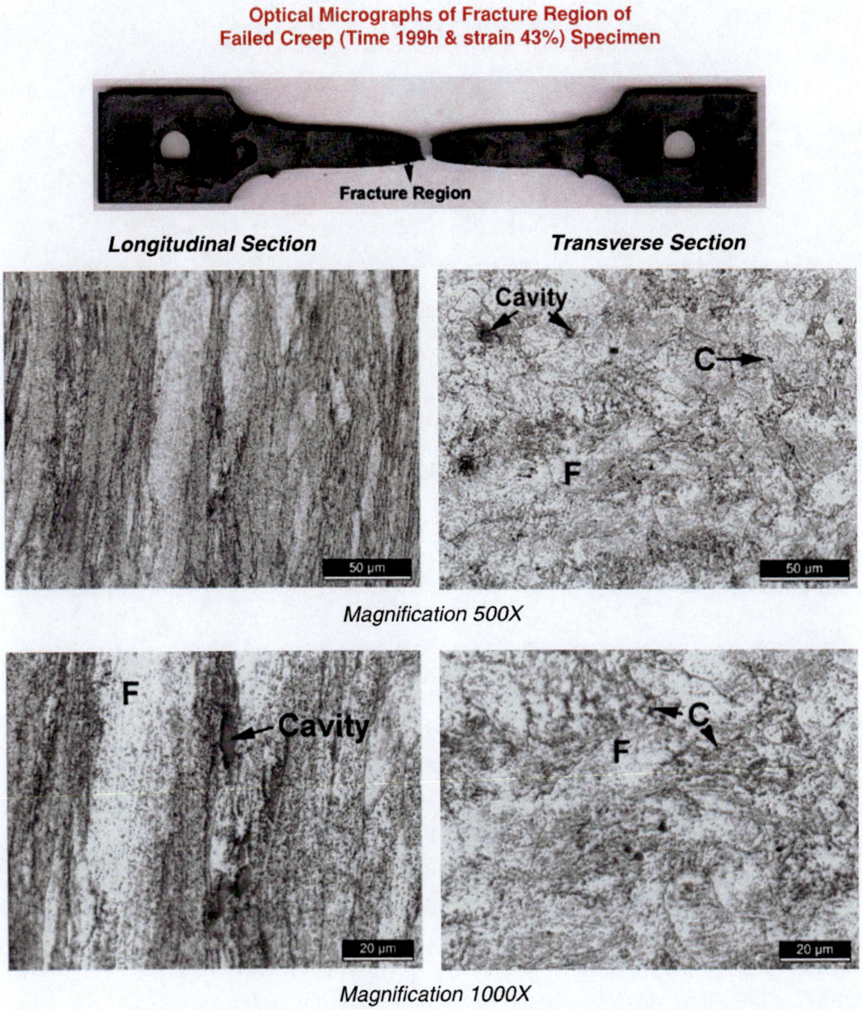

Fig. 4.76 Optical micrographs of **longitudinal and transverse sections** of P22 (2.25Cr–1Mo) steel after creep failed at 600 °C/100 MPa for 199 h, showing cavity formation in ferrite (F) matrix and carbide (C) precipitates at grain boundaries

SEM Micrographs of Fracture Region of
Failed Creep (Time 199h & strain 43%) Specimen

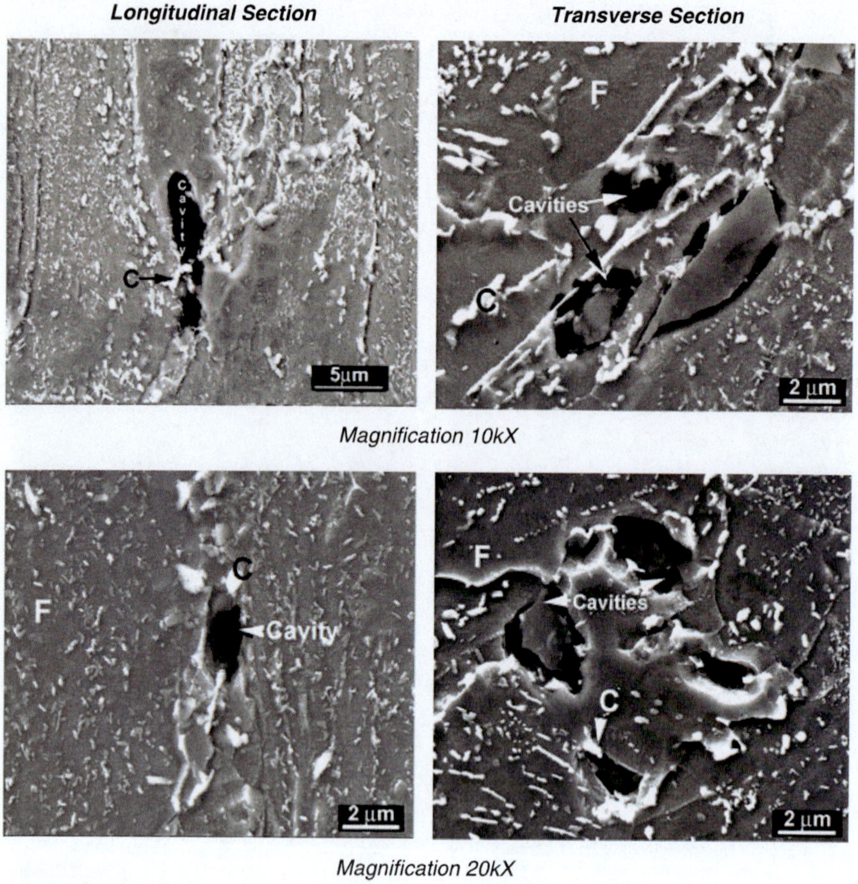

Fig. 4.77 SEM-SE micrographs of **longitudinal and transverse sections** of P22 (2.25Cr–1Mo) steel after <u>*creep failed*</u> at 600 °C/100 MPa for 199 h, showing the formation of cavities at carbide (C) precipitates in ferrite (F) matrix

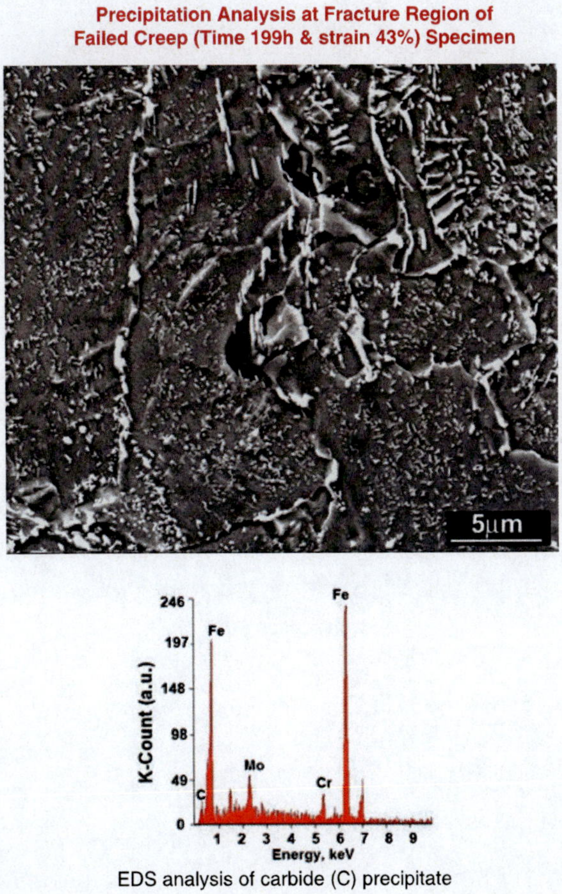

Fig. 4.78 SEM-SE micrograph of **transverse section** of P22 (2.25Cr–1Mo) steel after *creep failed* at 600 °C/100 MPa for 199 h, showing SEM-EDS analysis of (Fe, Cr, Mo)-rich carbide (C) precipitates at grain boundaries in ferrite (F) matrix

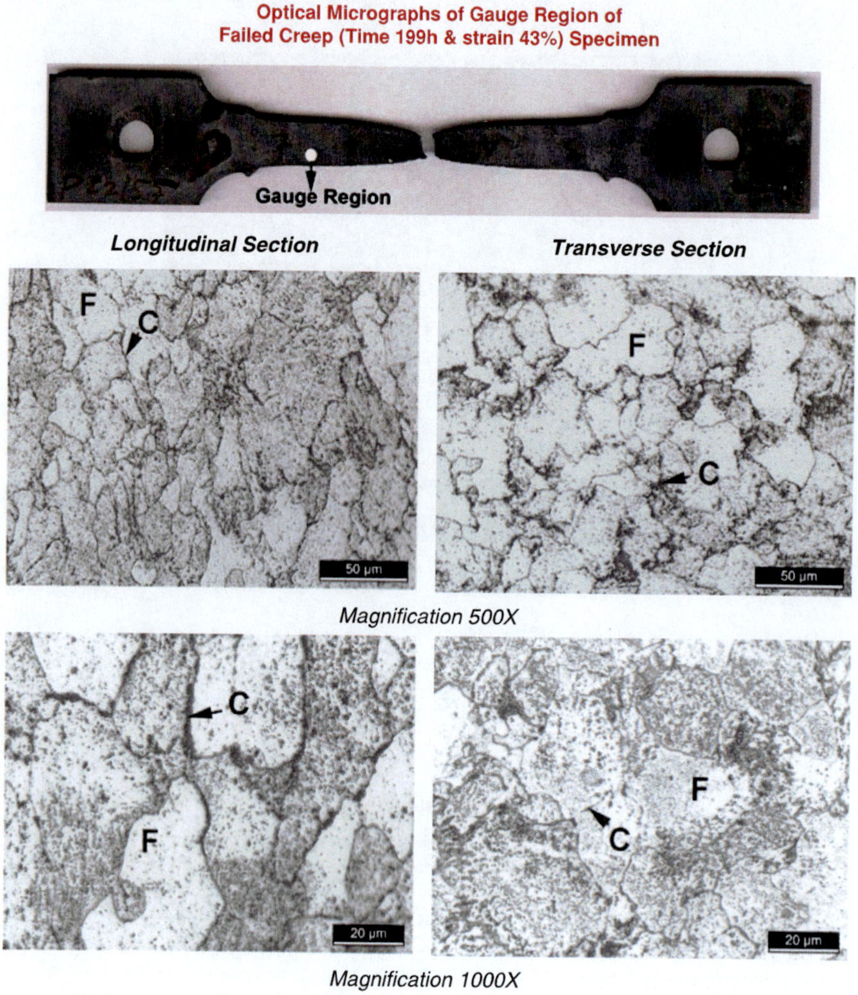

Fig. 4.79 Optical micrographs of longitudinal and transverse sections of P22 (2.25Cr–1Mo) steel after creep failed at 600 °C/100 MPa for 199 h, explaining carbide (C) precipitates at grain boundaries in ferrite (F) matrix

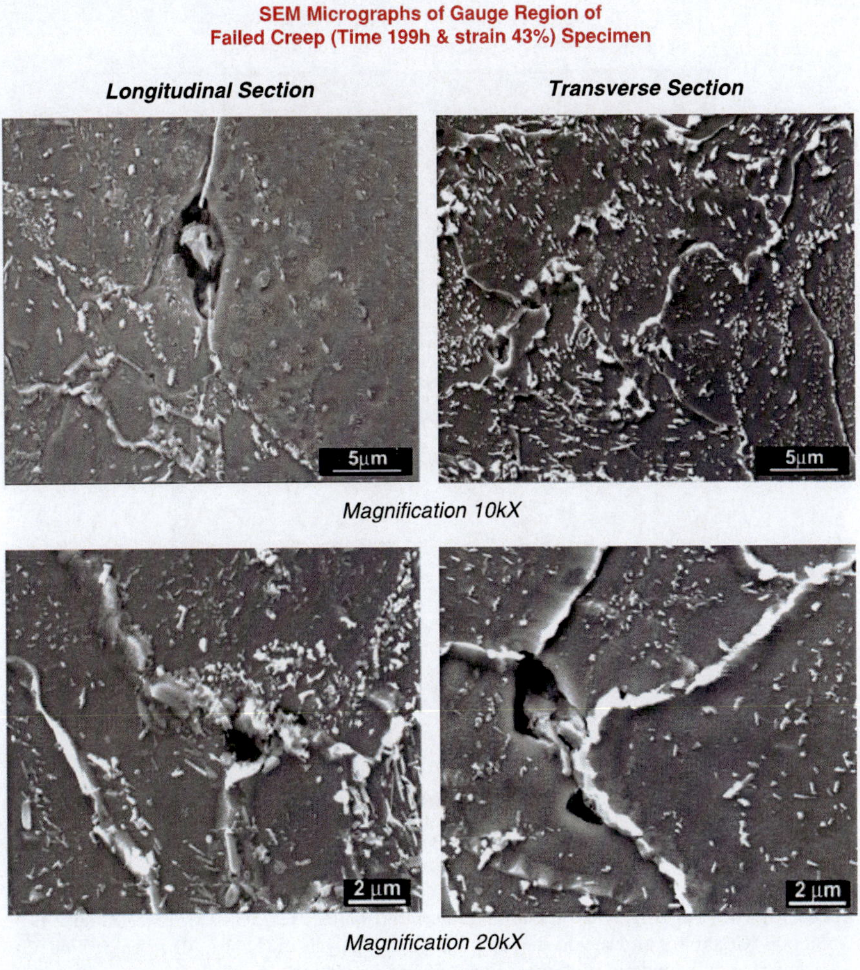

Fig. 4.80 SEM-SE micrographs of **longitudinal and transverse sections** of P22 (2.25Cr–1Mo) steel after <u>*creep failed*</u> at 600 °C/100 MPa for 199 h, showing cavity and carbide precipitates at grain boundaries and pearlite colonies and discrete pearlite in ferrite matrix

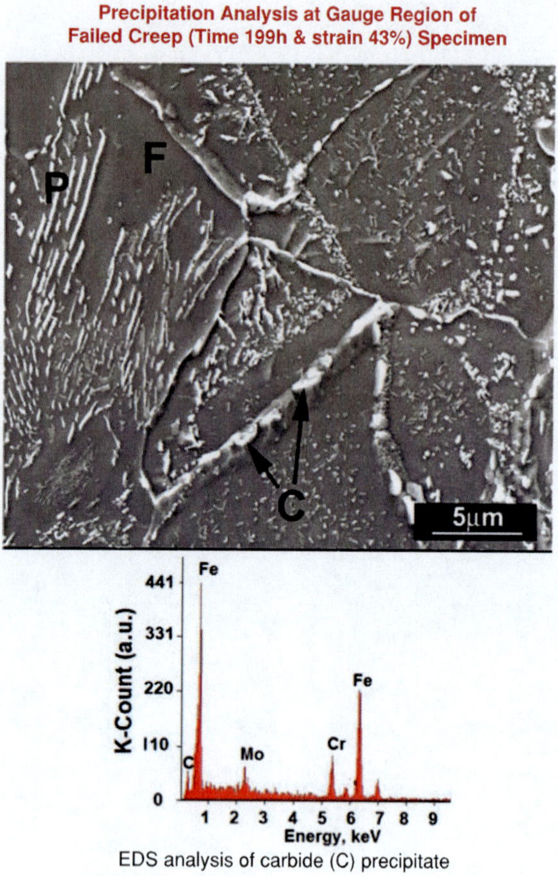

Fig. 4.81 SEM-SE micrographs of **transverse section** of P22 (2.25Cr–1Mo) steel after creep failed at 600 °C/100 MPa for 199 h, showing SEM-EDS analysis of (Fe, Cr, Mo)-rich carbide (C) precipitates at grain boundaries in ferrite (F) matrix and pearlite (P) lamellar

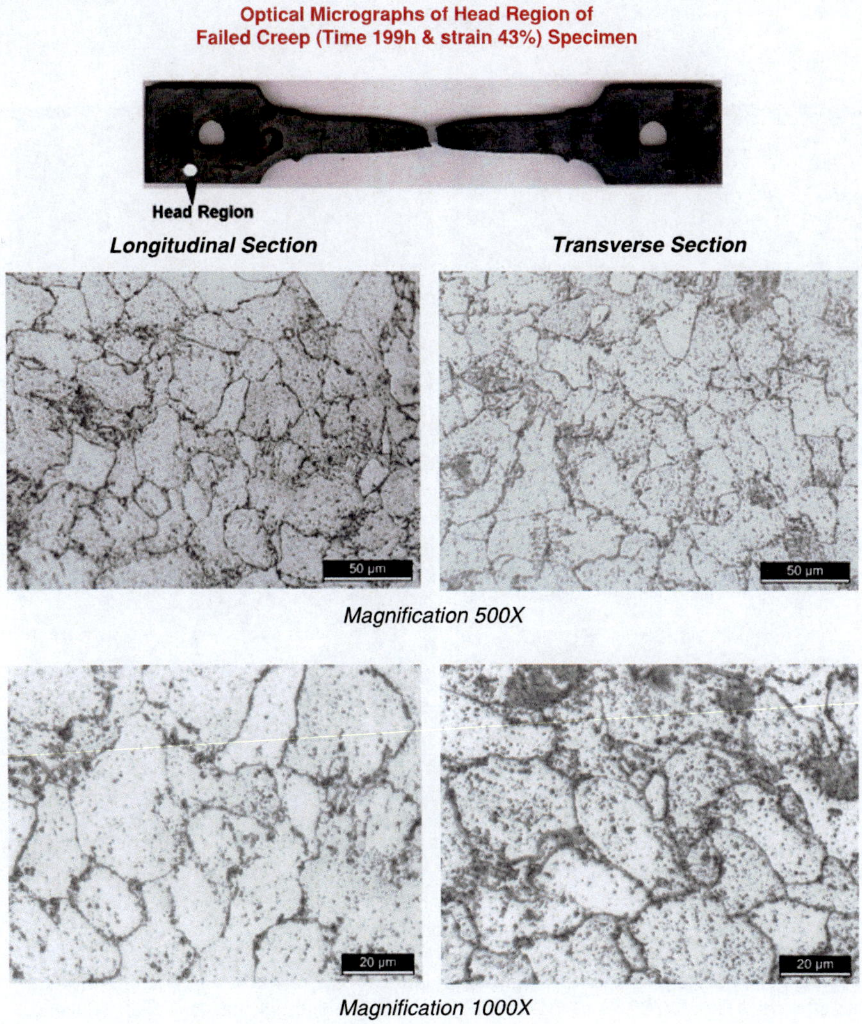

Fig. 4.82 Optical micrographs of **longitudinal and transverse sections** of P22 (2.25Cr–1Mo) steel after *creep failed* at 600 °C/100 MPa for 199 h, showing carbide (C) precipitates at grain boundaries and pearlite (P) in ferrite (F) matrix

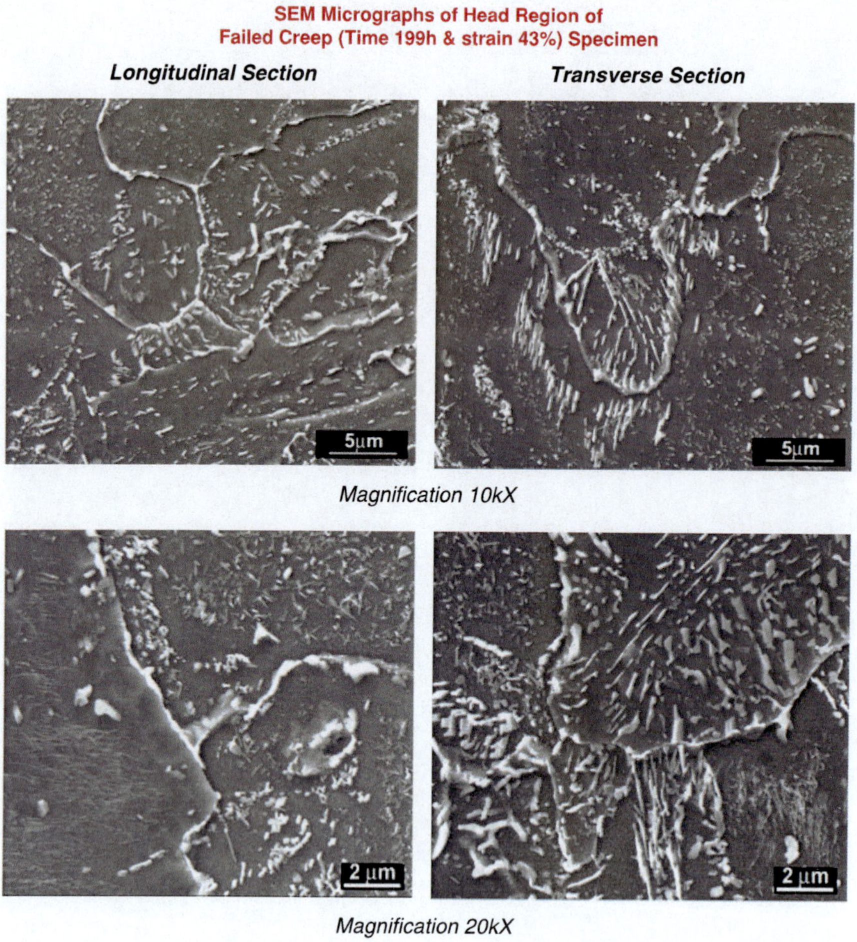

Fig. 4.83 SEM-SE micrographs of **longitudinal and transverse sections** of P22 (2.25Cr–1Mo) steel after *creep failed* at 600 °C/100 MPa for 199 h, showing carbide precipitates at grain boundaries and cementite (Fe_3C), pearlite colonies and discrete pearlite in ferrite matrix

Chapter 5
Effect of Creep Exposure on Material Behaviours

5.1 Precipitate Size Variation

See Fig. 5.1.

5.2 Hardness Variation

See Fig. 5.2.

5.3 Magnetic Coercivity and MBE Voltage

At initial creep deformation (t ~ 900 h), there are about no changes of coercivity up to ε_c ~ 6% and MBE voltage up to ε_c ~ 8%. The formation and growth of carbides (with Cr $\leq$ 15 wt%) occurs at initial creep deformation, which causes no change of magnetic properties. Subsequently, the generation of carbides of (Cr ~ 16–28 wt%) and (Cr > 28 wt%) leads to a 5–10% increase of coercivity, followed by a stagnant magnetic behaviour till ε_c ~ 30%. The magnetic MBE voltage is also rapidly increasing within this creep deformation. At higher creep strain beyond 30%, the coalescence and growth of carbides results in magnetic hardening, leading to a drastic drop of MBE voltage and rise of coercivity (Fig. 5.3).

Initially there is a gradual magnetic hardening (coercivity rise, MBE drop) below creep time of 1000 h and creep strain of 9%. This may be attributed to hindrance to magnetic domain wall movement due to precipitation of fine carbides. Beyond 1000 h, there is indication of magnetic softening (coercivity drop, MBE rise) up to 2000 h. In this zone, it is expected for the growth of carbide precipitates, decreasing the pinning density and thus favouring easy domain movement. Above 2000 h of creep time, there is a drastic magnetic hardening (coercivity rise, MBE drop). It is

R. K. Roy et al., *Microstructure Atlas of P22 Steel*,
https://doi.org/10.1007/978-981-99-0117-3_5

due to the precipitation of type-2 and type-3 carbides, leading to domain pinning and consequent magnetic hardening (Fig. 5.4).

Magnetic coercivity and MBE voltage of interrupted creep-exposed specimens of P22 at 575 °C/115 MPa show about no change in coercivity within creep strain of 10%. The small change of coercivity is due to a little carbide precipitation taking place at high creep temperature of 575 °C. However, MBE voltage varies within the ranges of 50–60% as it is a surface phenomenon. Above 17% of creep strain, there is a drastic magnetic hardening (coercivity rise, MBE drop), which is related to carbide growth and its domain pinning actions. At higher creep prior to failure, the cavity generation and growth of carbides lead to magnetic softening, indicating a drop in coercivity and rise in MBE. The high-temperature creep causes an erratic change in carbide precipitation nucleation and growth kinetics, leading to a large variation in magnetic behaviour (coercivity and MBE voltage) for each data point of creep-exposed sample. Accordingly, the deviation of magnetic parameters is higher for crept sample of 575 °C than that of 550 °C (Fig. 5.5).

The change in coercivity is very limited within 5% creep strain. The small change of coercivity is due to a little carbide precipitation taking place at high creep temperature of 600 °C. The high creep temperature results in carbides with mean diameter greater than 200 nm. Thus, the small change of magnetic parameters might be related to only the change of dislocation density. Like 575 °C, the 600 °C creep-exposed specimens also reveal a larger deviation of magnetic parameters than low-temperature (550 °C) creep-exposed one. It is due to an erratic change in nucleation and growth kinetics of carbide precipitation for high-temperature creep exposure (Fig. 5.6).

5.4 Nonlinear Ultrasonic (NLU) Parameter

See Figs. 5.7, 5.8, 5.9 and 5.10.

5.5 Conventional Ultrasonic Parameter

See Figs. 5.11, 5.12 and 5.13.

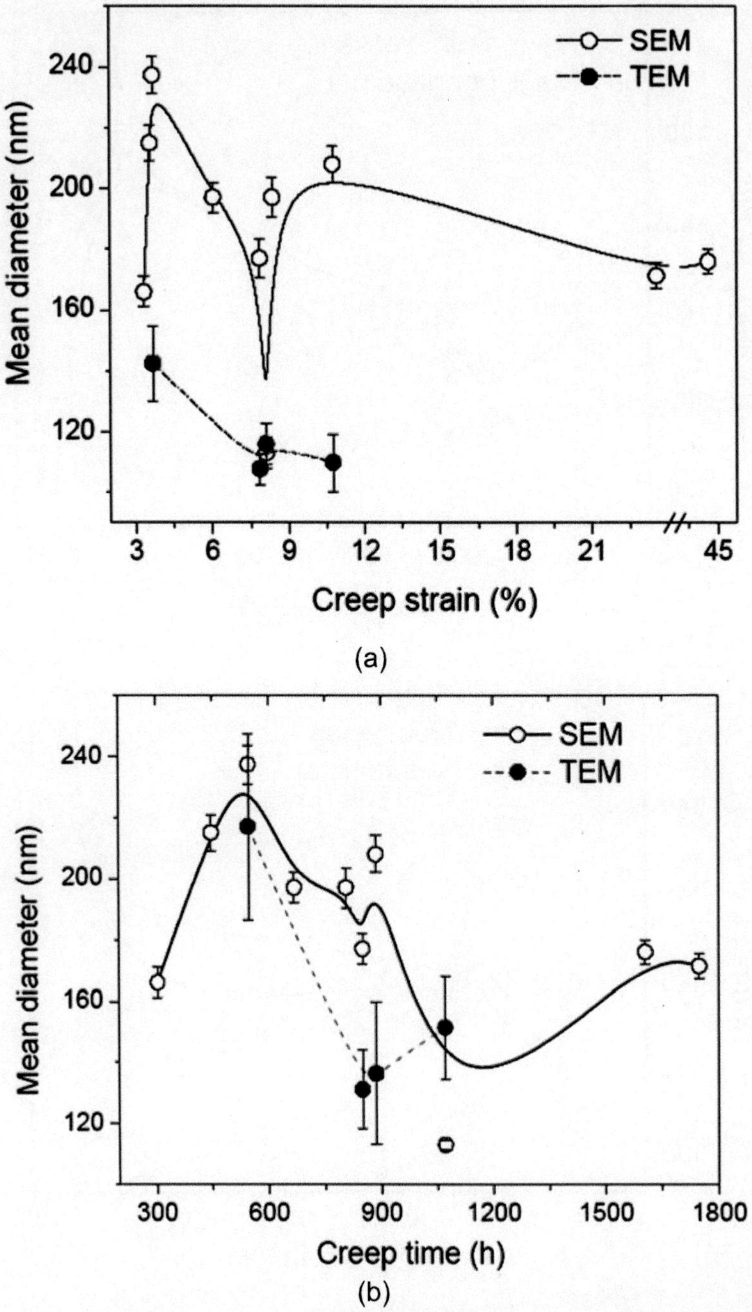

Fig. 5.1 Variation of mean diameter of carbide precipitates, derived from SEM and TEM, in P22 steel under creep exposed at 550 °C/115 MPa as a function of **a** creep strain and **b** creep time

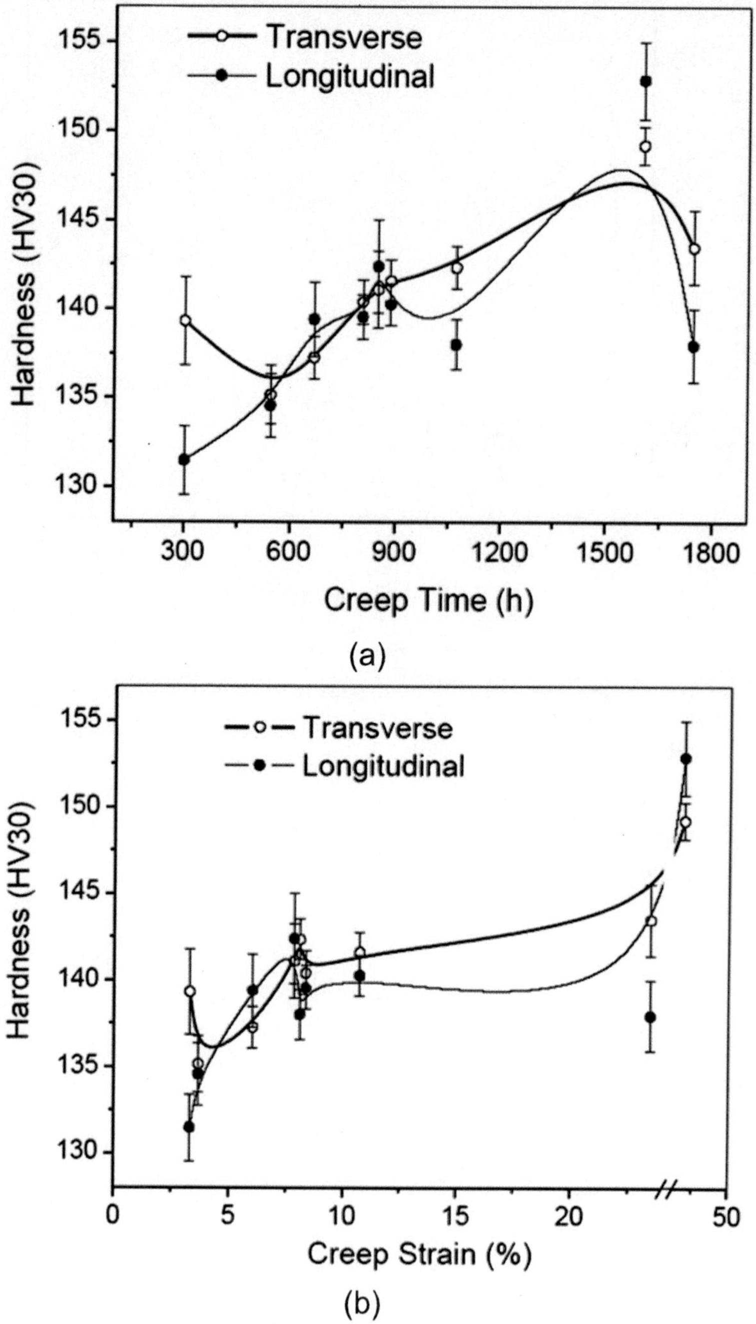

Fig. 5.2 Hardness variation of P22 samples under creep exposed of 550 °C/115 MPa as a function of **a** creep time and **b** strain

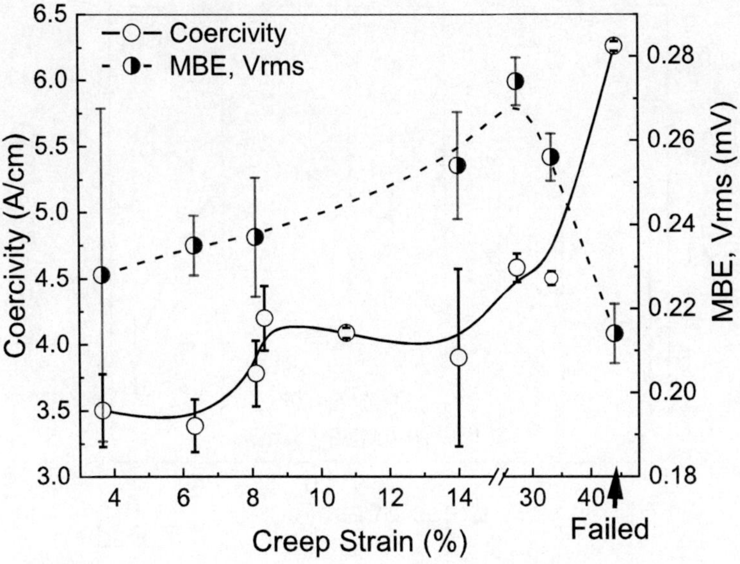

Fig. 5.3 Creep strain effect on coercivity and MBE voltage for different P22 steel multiple samples of single interruption (MSSI) under creep condition of 550 °C/115 MPa

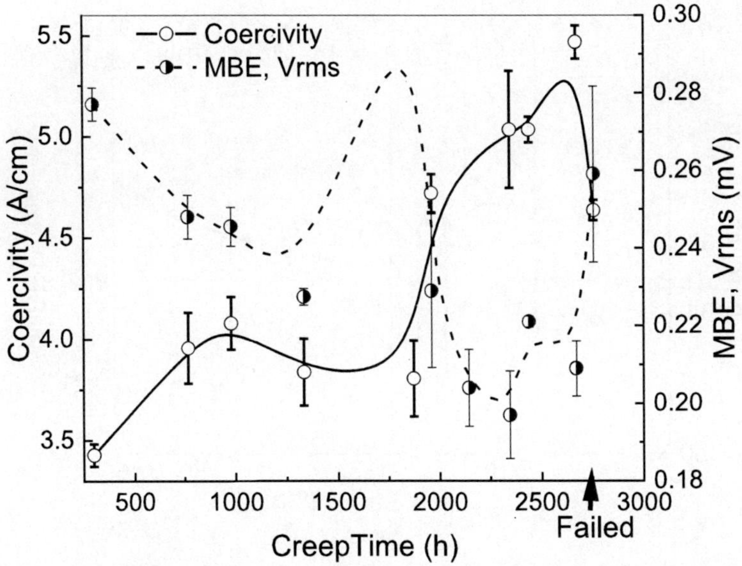

Fig. 5.4 Variation of coercivity and MBE voltage for a same P22 steel Single Sample under Multiple Interruption (SSMI) for creep condition of 550 °C/115 MPa as a function of creep time

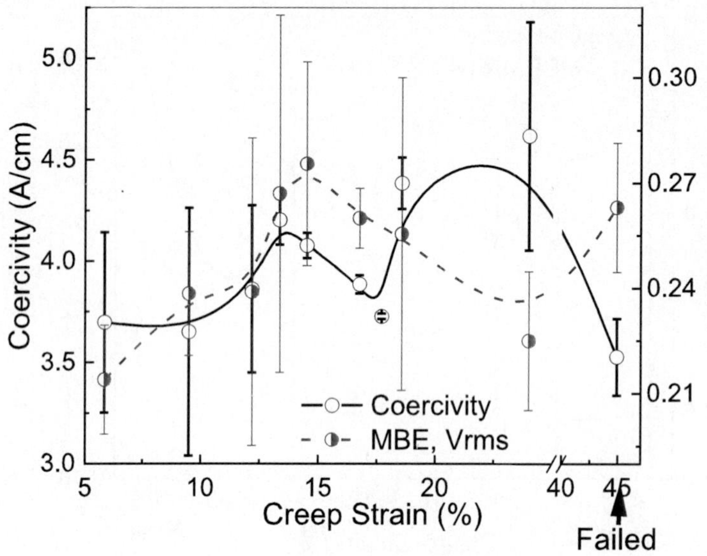

Fig. 5.5 Coercivity and MBE voltage of interrupted creep specimens P22 at 575 °C/115 MPa as a function of creep strain

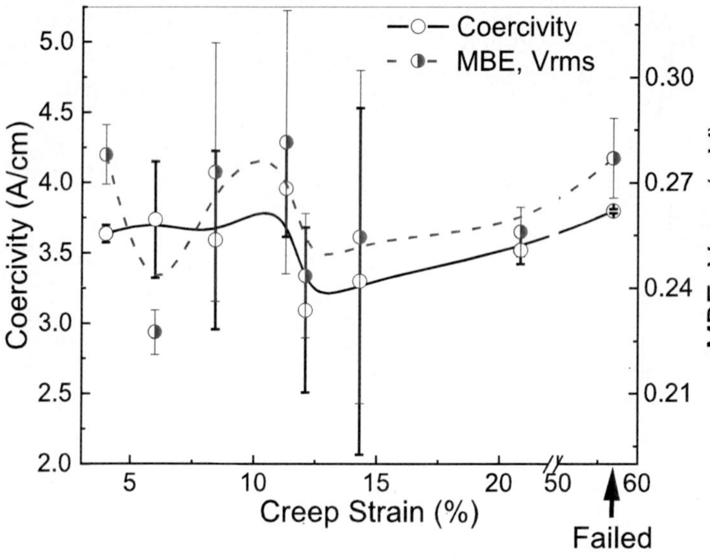

Fig. 5.6 Coercivity and MBE voltage of interrupted creep specimens P22 at 600 °C/100 MPa as a function of creep strain

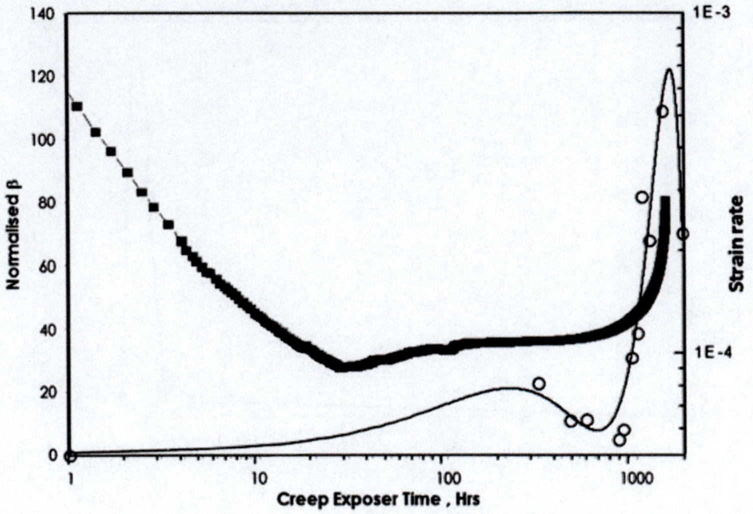

Fig. 5.7 Variation of normalized NLU parameter and strain rate with creep exposure time at 550 °C/115 MPa for P22 for MSSI samples

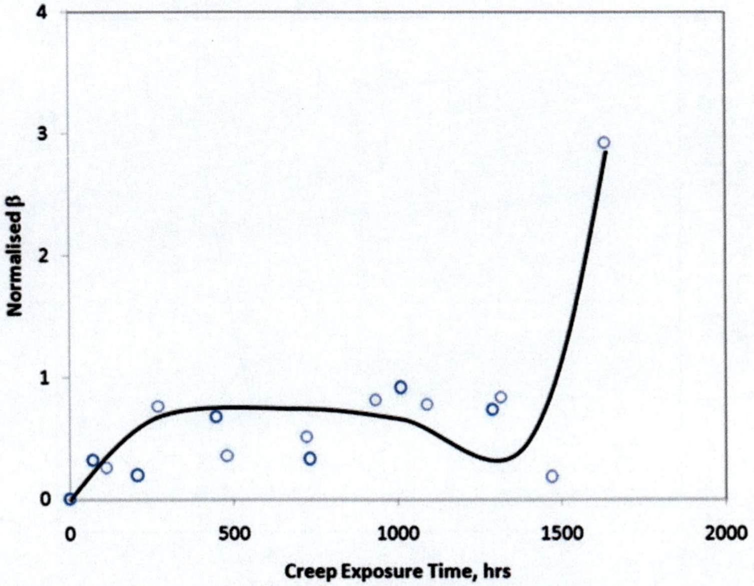

Fig. 5.8 Variation of normalized NLU parameter with creep exposure time at 550 °C/115 MPa for SSMI sample

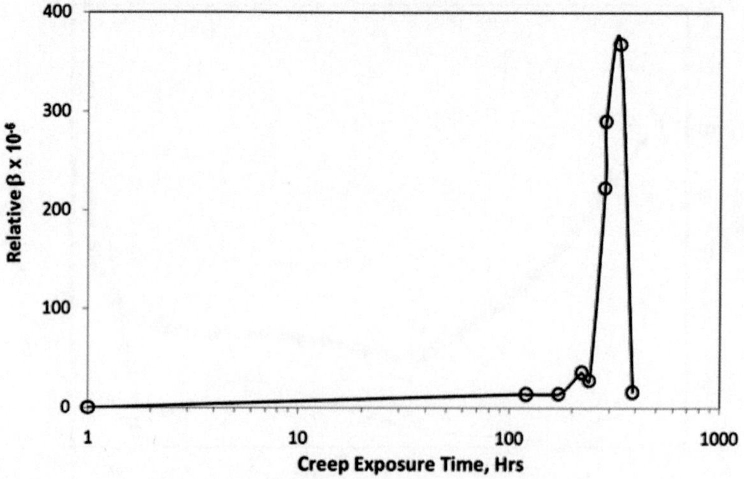

Fig. 5.9 Variation of normalized NLU parameter with creep exposure time at 575 °C/115 MPa for MSSI samples

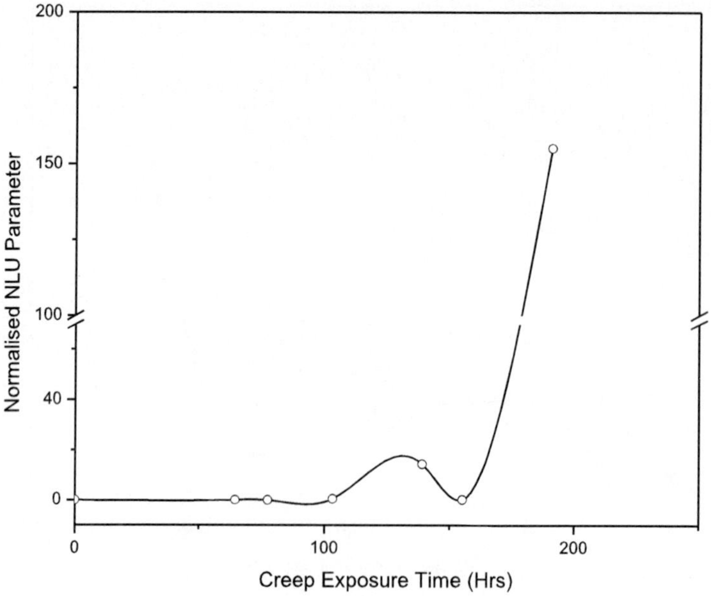

Fig. 5.10 Variation of normalized NLU parameter with creep exposure time at 600 °C/100 MPa for MSSI samples

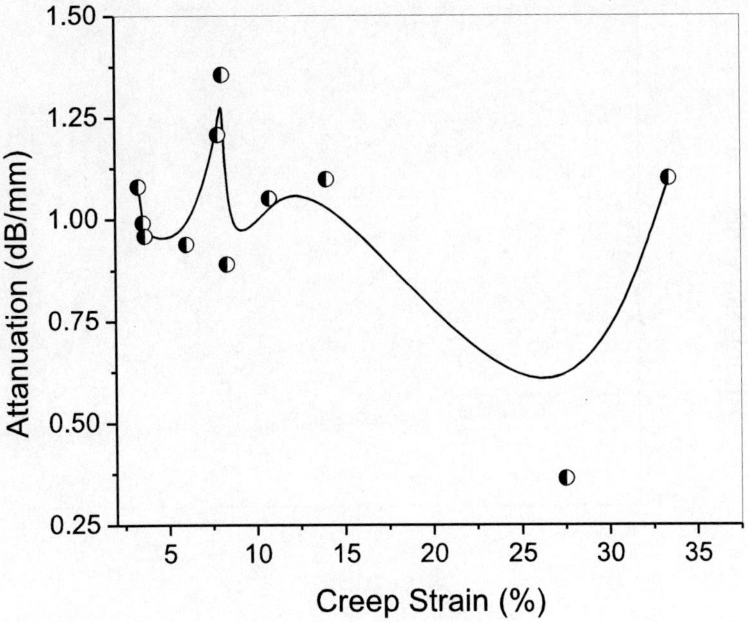

Fig. 5.11 Variation of ultrasonic parameter under 550 °C/115 MPa as a function of strain

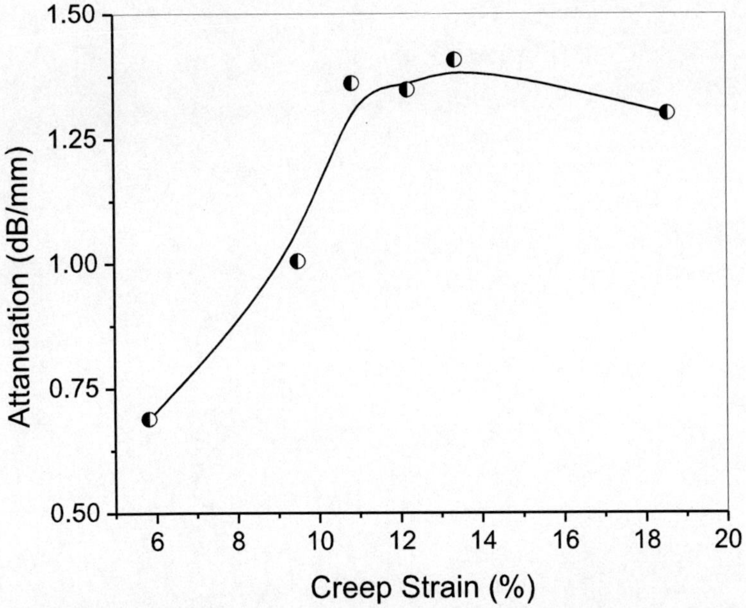

Fig. 5.12 Variation of ultrasonic parameter under 575 °C/115 MPa as a function of strain

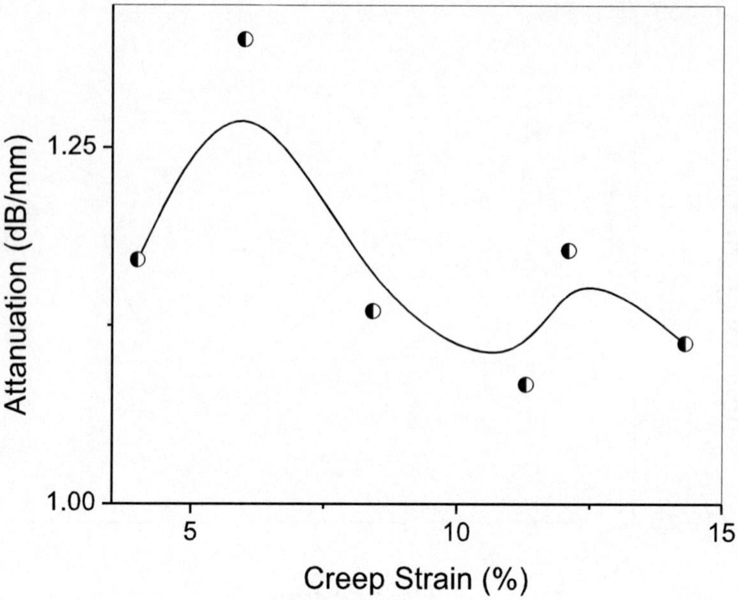

Fig. 5.13 Variation of ultrasonic parameters of P22 under 600 °C/100 MPa as a function of creep strain

Supplements

Brief Description of Experimental Methodologies

Sample Preparation

In this research, the creep specimen prepared along the longitudinal section of pipe and its dimensions are as per ASTM standard (Fig. A.1).

Creep Testing

The creep test was carried out with standard sample (Fig. A.1) in the constant load stress rupture test machine (Mayes make, model-MT3, 30 KN (max.), 950 °C (max.)), as shown in Fig. A.2. Two R-type thermocouples were tied on the specimen at the end of gauge length section. A preload of 5–10% of the target load was used and the samples were heated to the test set temperature within 1 h duration and soaked for 30 min for temperature equalization along the gauge length. The test temperature was controlled within ±2 °C of the set temperature. Rupture time was noted from the elapsed hour meter attached to the test machine. The displacement was measured using a pair of LVDT attached with the extensometer. Initial displacement readings were manually recorded during hot-loading and thereafter continuous recordings were done at every 15 min interval using a data-logger.

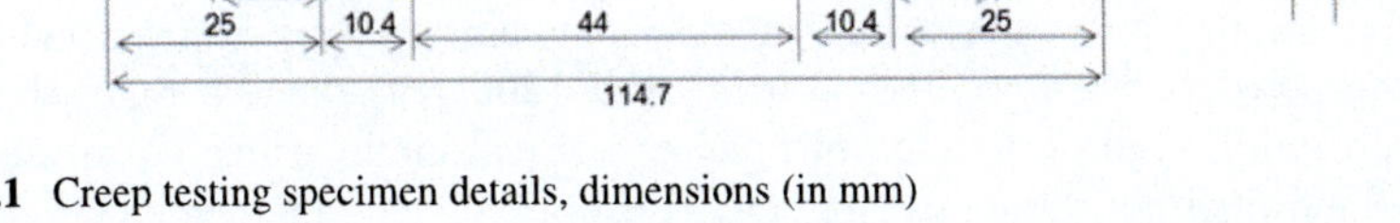

Fig. A.1 Creep testing specimen details, dimensions (in mm)

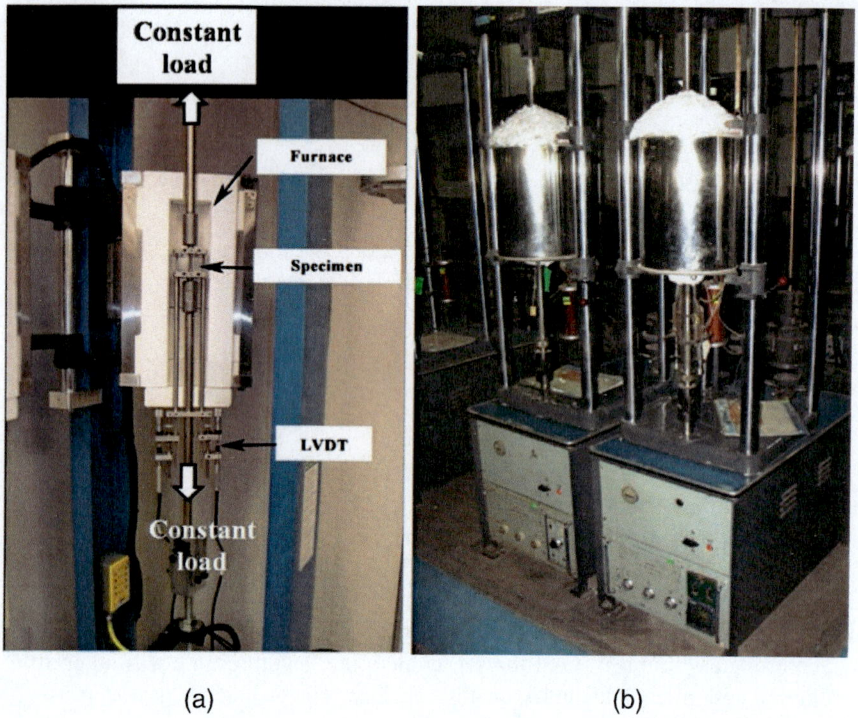

Fig. A.2 Creep testing machine, **a** sample before put into furnace, **b** sample packed in the furnace

Microscopy of Samples

Optical and Scanning Electron Microscopy

The samples of size $10 \times 5 \times 5$ mm^3 were cut along longitudinal and transverse (Fig. 2.1b) of pipe for as-received specimens and that of creep-exposed specimens at gauge and failed regions, as per requirement (Fig. A.3).

These specimens were suitably mounted by hot pressing using conductive mounting powders and were ground using 200–1200 grit silicon carbide abrasive papers. It is followed by coarse polishing on 10 μm cloths using 1 μm alumina solution and final polishing on 1 μm cloths using 0.05 μm alumina solution. The chemical etching was done by 2% Nital solution (2% nitric acid and 98% methanol). Chemically etched and dried specimens were examined under an optical microscope of Leica (Model: DM2500M) and scanning electron microscope of FEI FEG-SEM (Model: Nova NanoSEM430).

Optical microscopy (OM) was done under 200X, 500X and 1000X magnifications, and the images were captured by a digital camera. The selective images with high resolution are presented in this microstructure Atlas.

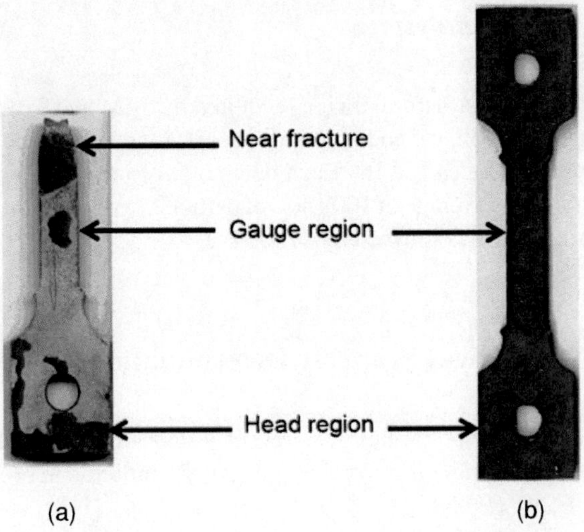

Fig. A.3 Microscopy of regions of creep-exposed specimens, **a** creep failed and **b** creep interrupted Multiple Samples and Single Interruption (MSSI) samples

Scanning electron microscopy (SEM) was operated at 15 kV and scanned at the lower magnification in order to extract the most representative microstructural features. The representative images are digitally captured at the magnifications of 5, 10 and 20 kX. The compositional elemental analysis of precipitates was done under 15 kV voltage through EDS analyzer attached with SEM.

Transmission Electron Microscopy

For transmission electron microscopy (TEM), the thin specimens of 0.3 mm thickness were cut from as-received and creep-exposed samples by electric discharge machining (EDM) with a proper cooling arrangement. These specimens were further thinned down by 1200–2500 grit silicon carbide abrasive papers till 100 μm fine foil which was finally punched to 3 mm diameter disc. The disc sample is again thinned down by twin jet polishing using solution of 4% perchloric acid in methanol at −40 °C and making the tiny holes, enabling the hole adjacent area to be electron transparent.

A Philips (Model: CM200) analytical transmission electron microscope operating at an accelerated voltage of 200 kV was used to identify carbides precipitates of as-received and creep-exposed samples. The precipitates were identified through selected area electron diffraction pattern (SAED) and elemental analyses by EDAX with TEM.

Image Analysis

The microstructural images acquired by SEM and TEM were analysed for the determination of size and area fraction of carbide precipitates by an image analysis software of Leica QWIN. Each data point of precipitate size and volume fraction was about the average of 1000 nos. of particles from 5 nos. of SEM images. The precipitate size is also confirmed by TEM.

Mechanical Property Determination

The tensile properties of as-received steel are determined by universal tensile testing machine under the crosshead speed of 5 mm/min at room and elevated temperatures of 550, 600 and 650 °C.

The hardness measurement of as-received and creep-exposed specimens was carried out by a diamond pyramid Vickers hardness tester of Chennai Metco (Economet VH50MD) at 30 kgf load with a dwell time of 10 s for indenting. The sample surface was ground up to 1200 grit SiC paper before hardness measurement. Each hardness value was the average of 5–10 nos. data points.

Brief Description of NDE Techniques

Electromagnetic Method

Among the electromagnetic techniques, Magnetic Barkhausen Emissions (MBE) and Magnetic Hysteresis Loop (MHL) have been used under this project. The magnetic hysteresis loop and the magnetic Barkhausen emissions signal are shown schematically in Fig. A.4 along with the measuring parameters. However, the coercivity (Hc) and the rms voltage (Vrms) are the most important parameters as regards with the microstructural changes are concerned. Hence, the emphasis has been given to measure those parameters. Both MHL and MBE parameters are to be measured by an electromagnetic sensing device, named *MagStar* (Fig. A.5), and developed jointly by CSIR-NML and M/s, Technofour, Pune.

For MBE measurement, a sinusoidal cyclic magnetic field of 239 A/cm (300 Oe) with an excitation frequency of 40 Hz was induced, and the Barkhausen signals were filtered with a wide band-pass filter (30–300 kHz). On the other hand, a sinusoidal cyclic magnetic field of 1200 A/cm (1500 Oe) with an excitation frequency of 50 MHz was induced for MHL measurement.

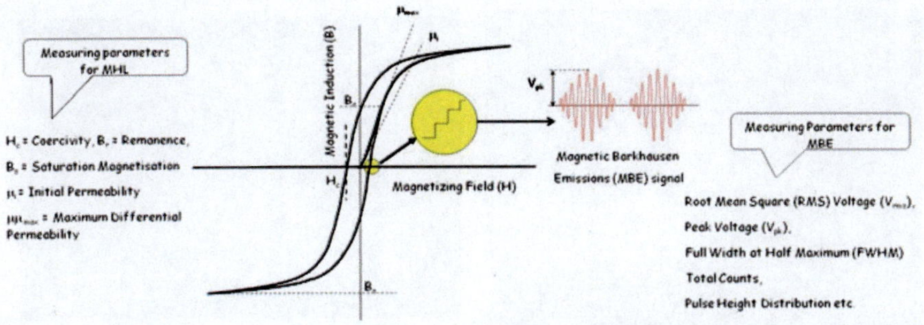

Fig. A.4 Schematic representation of Magnetic Hysteresis Loop (MHL) and Magnetic Barkhausen Emissions (MBE)

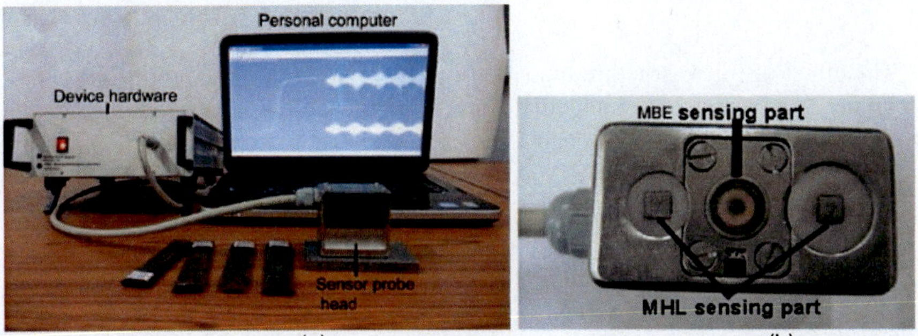

(a) (b)

Fig. A.5 **a** *MagStar* device, **b** combo sensor probe head for MHL and MBE measurement

Nonlinear Ultrasonic (NLU) Technique

The variation in the distribution of discontinuities due to material degradation results in a variation of nonlinear elasticity, and material degradation can be monitored by nonlinear ultrasonic (NLU) measurement. The most well-known NLU response is second harmonic generation caused by higher order nonlinear elasticity of the materials in question. The relation between arbitrary constants C_1, C_2 of nonlinear elasticity and two amplitudes of distorted ultrasound, fundamental (A_1) and second harmonic (A_2), was derived by Breazeal and Ford et al. Using fluid analogy and Lagrange's equation of motion:

$$\beta = \frac{C_1}{C_2} = \frac{8}{k^2 x}\frac{A_2}{A_1^2} \tag{A.1}$$

where K and x are wave number and propagation distance, respectively. The right-hand side of Eq. (A.1), which represents the ratio of primary and second-order terms of nonlinear stress–strain relation, is referred to as ultrasonic nonlinear parameter β.

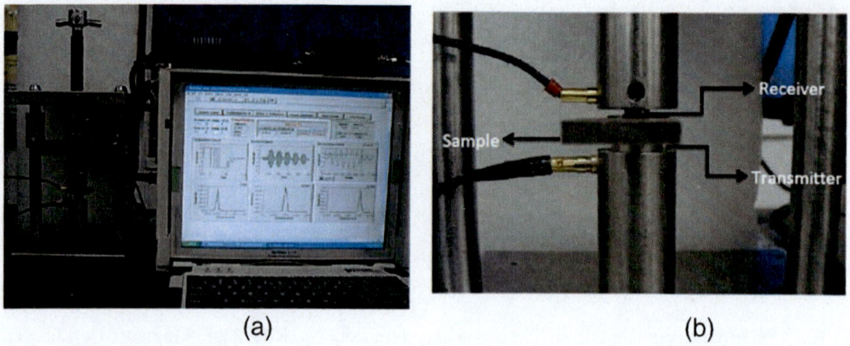

Fig. A.6 **a** "Ultraβ": Portable system for NLU parameter measurement, **b** sample holder for applying constant load to the sample

The effect of microstructure transition by degradation or material processing to β variation has been widely studied, i.e. precipitation hardening heat treatment, cold work, isothermal ageing, creep and fatigue, and has been reported that dislocation density, precipitation distribution, micro-void and externally applied or residual stress are primary factors to ultrasonic nonlinear parameter β.

A portable instrument "Ultraβ" designed and developed at CSIR-NML was used to measure β (Fig. A.6a). A sample holder is also designed to place the sample in between the transmitter and receiver probes under constant load (Fig. A.6b).

For the measurement of NLU parameter, a 5 MHz narrowband transmitter and a 10 MHz broadband receiver were used. NLU measurements were carried out in each sample in the as-received condition prior to creep testing (β_{zero}) and after interruption (β_{damaged}). The normalized β was defined as mentioned in Eq. (A.2)

$$\beta = \frac{\beta_{\text{damaged}} - \beta_{\text{zero}}}{\beta_{\text{zero}}} \quad \text{(A.2)}$$

A 5-cycle tone burst signal was used to excite the transmitter at two different voltages, 1050 and 1200 V. The received signal was analysed using *Cal*β © software for online NLU parameter determination after calibration.